THE SPEEDY VEGETABLE GARDEN

Timber Press
London • Portland

Page 2: *Fresh, crunchy and nutritious just-sprouted peas, ready to eat after just a few days.*

Photographs by Mark Diacono
Illustrations by James Nunn

Published in 2013 by Timber Press, Inc.

The Haseltine Building
133 S.W. Second Avenue, Suite 450
Portland, Oregon 97204-3527
timberpress.com

2 The Quadrant
135 Salusbury Road
London NW6 6RJ
timberpress.co.uk

Printed in China
Book design by James Nunn

Library of Congress Cataloging-in-Publication Data
Diacono, Mark.
The speedy vegetable garden / Mark Diacono and Lia Leendertz. – 1st ed.
p. cm.
Includes index.
ISBN 978-1-60469-326-3
1. Vegetable gardening. 2. Cooking (Vegetables) I. Leendertz, Lia.
II. Title.
SB321.D53 2013
635–dc23 2012021278

A catalog record for this book is also available from the British Library.

Contents

Introduction

The traditional perception of gardening is that it's all about the long view. A gardener must be stoic and long-suffering: seeds sown in spring will come to fruition in autumn. We sow, we tend, we wait, and if things go our way we finally harvest. But while it's true that a great many delicious and worthwhile plants require such forbearance, there are also a large number that suit even the most impatient gardener. These are the crops that will be ready to eat in weeks, days, and even hours, and they are the subject of this book.

This isn't about cheating: it's about plants that are actually at their best grown quickly and impatiently. Some are small salad leaves and tiny micro greens that must be enjoyed young and fresh – be slow to harvest and you miss their best moment. These are fresh, lively and zingy flavours, flavours that can either fade or become bitter and overly strong as the plant grows on towards maturity.

Of course, 'fast' is relative. While there are crops in this book that can be on your plate within a few days of sowing, others may take many weeks. What we have sought out are the quickest types and varieties of each crop: for example, cherry tomatoes ripen faster than beefsteak tomatoes because there's less bulk to ripen, and new potatoes are faster to the saucepan than maincrops because they are picked when little, soft and sweet, before their skins have hardened.

Pick radishes while they are small and all crunch, and before any woodiness creeps in.

We have also chosen plants that aren't traditionally quick to harvest but are actually improved by being picked when immature, such as carrots and beets that are more sweet and tender when picked at half or even a quarter of the size you would find them in the shops. Many of these crops – the edible flowers, the delicate but punchy micro greens and the tiny, sweet vegetables – are hard to find in the shops, and are a daunting price when they do appear.

Edible flowers and micro greens are quick to grow and make for lively and colourful salads.

The one thing you will miss out on with speedy growing is bulk, but what you will get in return is layers of flavour: a sprinkle of hot and peppery micro-green radish here, a sweet and nutty, barely cooked new potato there, a garnish of cucumbery borage flowers to finish a dish. These are the crops that will mark out your cooking as distinctly and unquestionably homegrown.

These quick crops are a great place to start if you are a beginner and want to get a few fast and easy crops under your belt to set you up with a little confidence. They are also a great place to graduate to if you are an experienced gardener who specializes in those crops that are big on bulk but mellower of flavour, and you're craving some of the exciting stuff. This is the quick, easy and tasty end of growing your own. There is no suffering here.

Mark and Lia

Soaks and Sprouts

Something happens when you soak a seed overnight – quite a few things, in fact. Protein, vitamin and digestible energy levels all surge as the seed kicks into life. Metabolic activity increases. The dry, starchy resting state becomes active and vital, primed for life.

At these early stages the seed is furiously generating all the complex materials it needs to launch itself into the world. It seems almost cruel to cut it off at this hopeful, vulnerable stage, but if we catch it and eat it now a little of that vitality can be ours. In the germinating process all of the goodness in the seed becomes more 'bioavailable', which essentially means it is easier to digest. Germination activates a plant's stored energy, so a sprouted seed has fewer calories yet more vitamins and minerals than the dry version. This is a superfood that you can grow on your kitchen table.

In fact, even a sprouted seed isn't the very fastest thing you can grow to eat. Strictly speaking, that honour should go to the 'soaks' – nuts and seeds that haven't yet produced a sprout, but have soaked just long enough to swell up and think about germinating.

So this is why the soaks and sprouts make up the first chapter of the book; if you are the truly impatient gardener that this book is aimed at, here is the place to start. You'll have results in a matter of days, or hours in the case of the soaks.

Scatter crunchy, sweet mung bean sprouts over salads and into stir-fries.

This isn't all just worthy and healthy eating, with no pleasure to speak of. Sprouts are a delicious ingredient, crunchy and fresh and with far more variety and complexity than a brush with the ubiquitous (though admittedly sweet and lovely) mung bean sprout would suggest. Sprinkle them with abandon to bring a fresh, nutty taste to sandwiches and salads.

Lia: "Soaks are a revelation – do give them a try. Soaked almonds remind me of times I have eaten nuts straight from the tree, while they're still juicy and fresh. Just the act of starting them into life with a quick soak reinvigorates them."

TECHNIQUES

It's incredibly easy to get a seed to sprout: just give them moisture and warmth and off they go. It's what they are designed to do. They also need plenty of air, and most sprouting equipment allows quick drainage as well as easy watering. Most won't need light so they can stay in a dark corner, but if after a few days they start to show small leaves, you can move them into the light so that the leaves will quickly turn green. Never start them off in direct sunlight as this can heat them up and dry them out, and indeed an afternoon in the sun can even cook them once they have sprouted. Average room temperatures with no dramatic fluctuations are perfect for sprouting seeds.

It's possible to sprout seeds in a jam jar with a small square of muslin secured over the top with an elastic band, but it's not ideal as the seeds are a little crammed together and drainage isn't great, leaving scope for moulds to develop. For large seeds, use a sprouting bag and for small ones a tray sprouter. Both are available from specialist suppliers, and thus equipped you can sprout the whole range of seeds.

SEEDS FOR SOAKING AND SPROUTING

The sprouting times given below are merely a guide – in warmer weather the sprouts will grow faster, in cooler, slower. Make a habit of tasting your sprouts at every stage of the process after the initial soaking. Many are sweeter when they are younger, and you may find you prefer to grow them for a shorter or longer time than is suggested here.

Some sprouted seeds will still have a fairly large hull attached. To remove them, put the sprouts in a large bowlful of cold water and gently agitate them with your fingers. The hulls will float to the top, where you can skim them off.

Sprouted seeds will store for a couple of days if kept in the fridge in a sealed container. They will keep best if the surfaces have been allowed to dry off, so leave them a good few hours after their final rinse before putting them into the fridge. Really, though, this isn't a crop you want hanging around. If you eat them as soon as they're ready you'll be enjoying them at their best.

Using a sprouting bag

Big seeds such as beans and chickpeas do well in a sprouting bag. Bags take up little room and are easy to rinse: you simply run water into them and leave them to drain.

Using a tray sprouter

Small and delicate seeds such as fenugreek would find the going a bit rough in a bag. For these seeds, use a sprouter with a series of draining trays. The seeds sit on the top and are easily rinsed. Most have several trays so you can sprout lots of different seeds at once.

Sprouting bag

Soak your big seeds for 8–12 hours. They should swell up.

Moisten the bag, pour the seeds in and give them a quick rinse through.

Rinse every 12 hours by running the bag under the tap, and leave to drip dry. Move them around at each rinsing, so they don't root into the fabric.

When the seeds sprout, eat them straight away after a final rinse. Store any left over in the fridge.

Tray sprouter

First soak your seeds for 8–12 hours. They should swell up.

Spread out the seeds on the surface of the sprouting tray. Remember that they are going to grow considerably in a short amount of time, so leave plenty of space around them.

Rinse the seeds every 12 hours. The roots will grow slightly into the drainage holes and fix the sprouts in place. Once this has happened you can submerge the whole tray for a minute and then drain.

When the sprouts are long enough, rinse, then eat or store in the fridge.

Pumpkin

4 hours

This is the quickest thing you will find to 'grow' in this book, though there is so little to do that term isn't really warranted. Still, by just starting pumpkin seeds into growth you will transform them into softer, more buttery seeds, with a less brittle texture. They are primed with nutrients, particularly potassium, phosphorus and vitamins A, B, C and D.

Cultivation

To treat pumpkin seeds as 'soaks', rinse a handful of the seeds first in case they are dusty. Soak in plenty of water (one part seeds to three parts water) for 1–4 hours. You can leave for longer, when you will start to see the tiny nub of the nascent sprout, but there is really no need. They are great after just an hour.

Harvesting and eating

Soaked pumpkin seeds don't store well, so make only as many as you need and eat them right away, after a quick rinse. Sprinkle them on salads and in sandwiches.

Peanut

3 days

Peanuts are eaten as soaks, sprouted only to the point where the root shows as a tiny bulge at the tip and the surface starts to split open. They take on a different texture as they swell, and a different sort of crunch: moist and fresh, almost vegetable-like, rather than dry and brittle. They are packed with vitamins A, B, C and D in this primed state, and are a source of calcium, iron and magnesium.

Cultivation

Soak peanuts for up to 12 hours and then sprout them for up to three days, rinsing every 12 hours.

Harvesting and eating

Soaked peanuts store fairly well for about a week in the fridge. Use them anywhere you might add unsoaked peanuts, such as in a stir-fry. Try them alone as a snack: they are delicious.

Nutty sprout coleslaw

Sprouted nuts and seeds are great in coleslaw, giving it substance and depth while matching the crunch of the cabbage and carrot perfectly.

3 carrots, grated
½ small white cabbage, finely shredded
handful of pumpkin soaks
handful of peanut soaks
4 tablespoons mayonnaise

Combine all of the ingredients in a large bowl and mix together well. Serve straight away.

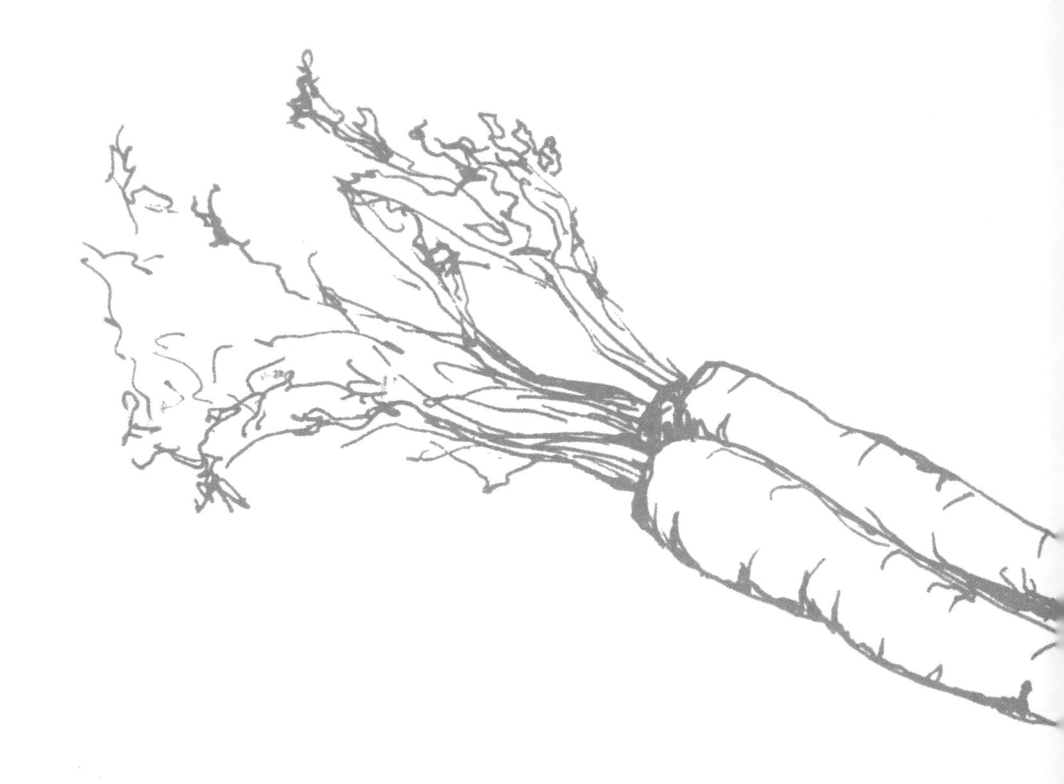

Almond

3 days

While soaked almonds are expensive compared to other soaked or sprouted seeds, they are a real luxury, reminiscent of almonds picked fresh from the tree. Less intensely nutty in flavour than a dried nut, they have a soft, buttery texture. They are ready quickly, and being in the throes of the very early stages of germination they are packed with amino acids.

Cultivation

They are ready after soaking for 12 hours. You can also try sprouting them for up to three days, rinsing every 12 hours.

Harvesting and eating

Almond soaks will keep for up to a week in the fridge.

Lia: "Keep a bowl of almond soaks to hand as a snack. They are lovely with dried apricots."

Labneh with soaked almonds and honey

Labneh is Greek yogurt, strained until it forms a solid consistency halfway between soft cheese and yogurt. It is often used in savoury dishes, but is just as good as a dessert. You can just use Greek yogurt straight from the tub, but it won't be as rich.

Serves 4

500g (17 ounces) labneh or Greek yogurt
handful of almond soaks
drizzle of runny honey
drizzle of extra virgin olive oil

1 Twenty-four hours before you want to eat the dish, lay a piece of fine muslin over a large bowl and spoon in the yogurt. Gather the corners of the muslin together and hang the yogurt above the bowl (from a kitchen cupboard door handle or the struts of an upturned stool, for instance), leaving it to drip. It will form into a round, sliceable cake.

2 Lay a slice on each plate and sprinkle soaked almonds on top. Drizzle with honey and then with a few drops of the olive oil.

Sunflower

4–6 days

Sunflowers make a substantial pure white sprout, crunchy and fresh, not unlike the mung bean sprouts you can buy in the shops. Grow them in the sun, where they will green up and be even more nutritious, or keep them in the dark for sweeter, pure white shoots. They are high in amino acids, potassium, phosphorus and magnesium.

Cultivation

Soak for 12 hours and then sprout for 4–6 days, rinsing every 12 hours. If you would like them to green up, move them onto a windowsill once they have sprouted. Sunflower sprouts are good at any stage of sprouting, and if you can get hold of hulled seeds you can even use them as soaks: they will need soaking for just an hour or two and are possibly even tastier than the sprouts, sweet and crunchy.

Harvesting and eating

Sunflower sprouts will keep for a couple of weeks in the fridge, but alternatively just let them keep on growing in the sprouter, as their flavour will develop and change as they grow. They are useful for adding cool crunch to stir-fries and salads.

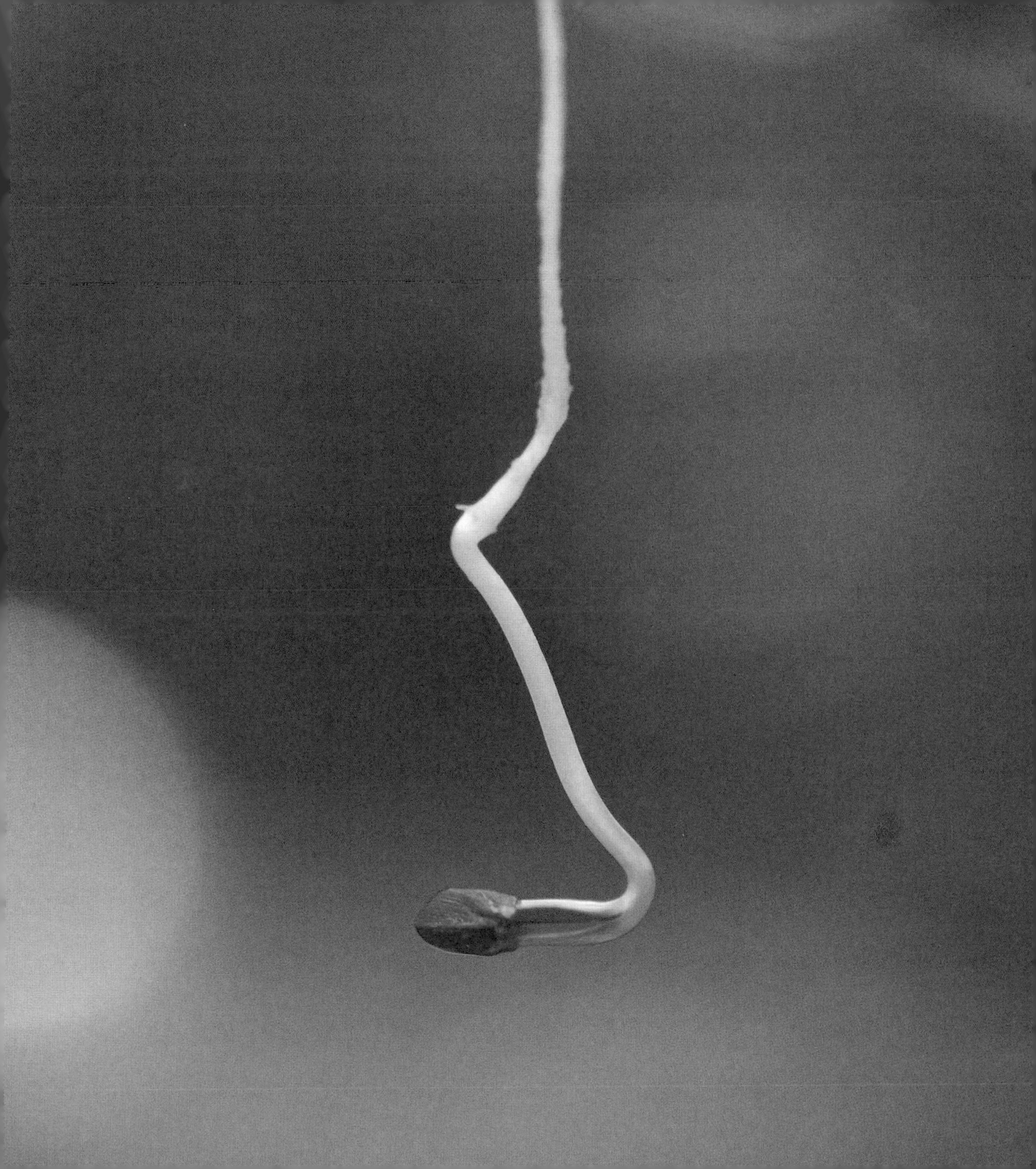

Radish

3–6 days

If you think radishes are hot and peppery, try a radish sprout – all that fire is concentrated into a tiny little seedling, making this an explosive addition to salads. In fact these are not the seeds of the little round red salad radishes, but of big, hot daikon types, widely grown throughout the Far East for the crisp flesh that holds its texture throughout long stewing in soups. The flavour is vital and invigorating, and it is not surprising to find that the sprouts are packed with calcium, magnesium, iron, carotenoids, amino acids and vitamins. They certainly taste as if they are full of goodness. These are particularly beautiful sprouts, pure white tinged with purple and with tiny green leaves.

Cultivation

The seeds tend to float, so push them below the surface with a finger when soaking them. Leave them for 8–12 hours then grow on for 3–6 days, rinsing every 12 hours.

Harvesting and eating

Radish sprouts will keep for up to two weeks in the fridge. They are perfect for pepping up a salad of mild lettuce leaves, or for sprinkling onto clear oriental soups such as dashi.

Lia: "All sprouts will green up a little if placed in gentle indirect light for their last day of growing and this will give them an extra boost of nutrients. They tend to be sweeter before greening up, and I prefer them pale. Try them both ways to find your own preference."

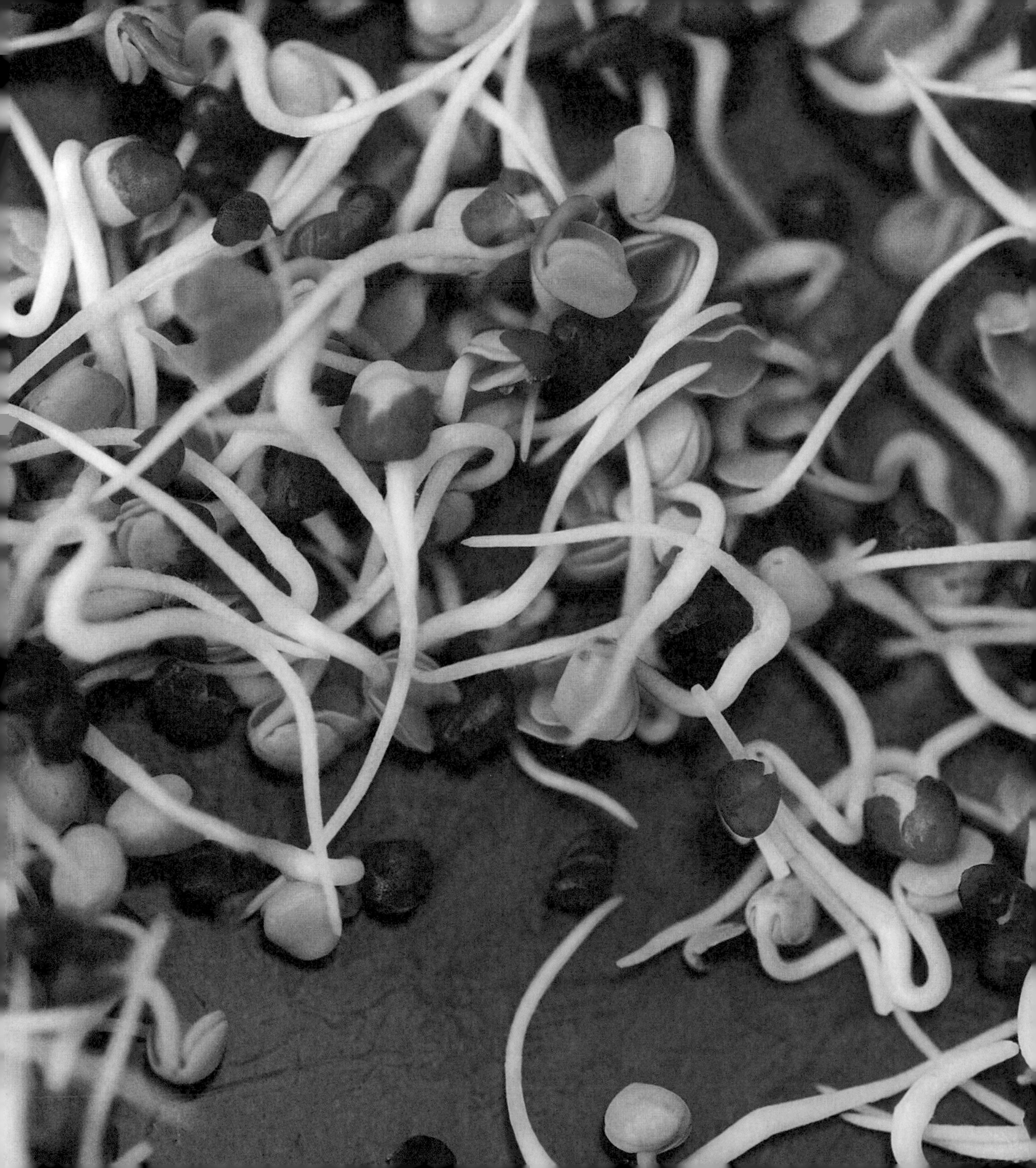

Mustard

5 days

Even the tiniest mustard sprout will make its presence felt, for these are strong, spicy and aromatic, much like the familiar seeds and condiments but fresher and sweeter, with a powerful horseradish flavour. Mustard sprouts are rich in carotene, amino acids and trace elements, as well as being packed with vitamins.

Cultivation

The seeds float, so push them under when soaking to encourage them to sink. Soak for 8–12 hours and sprout for around five days, rinsing every 12 hours.

Harvesting and eating

Mustard sprouts don't store well – you can keep them in the fridge for a few days, but they are best fresh. Just as mustard seeds ground into a sauce match beautifully with cold meats, try these sprinkled on thinly sliced ham or beef and other cold cuts.

Mark: "These sprouts are assertive enough to hold their own against other strong tastes. I stuff them into pitta breads alongside lamb."

Clover

6 days

Clover acts as a 'base sprout' for stronger flavours, being nutty but mild. All sprouts are nutritional dynamos but these have the highest number of isoflavones, which have powerful anti-cancer properties, and are particularly rich in vitamins and minerals. Small and delicate, they are great used en masse for crunch.

Cultivation

Soak for 8–12 hours and then sprout for five days, rinsing every 12 hours.

Harvesting and eating

Clover keeps quite well in the fridge for at least a week or two. Use as the base for a sprouting mix, with smaller amounts of stronger-flavoured sprouts.

Mung bean

3–5 days

When people say 'bean sprouts', mung bean sprouts are generally what they mean. The most widely eaten sprout on the planet and the sprout that started it all, mung bean sprouts are juicy, sweet and nutty, with a fabulous crunch. Though good in themselves they are a great carrier of flavours and are widely used in a great range of Asian dishes.

Cultivation

It's impossible to re-create the large, fat-rooted sprouts you can buy in supermarkets in your home sprouter: these are grown in huge vats with the use of chemicals and gases. You have two options. You can grow mung bean sprouts in the same way as all other sprouts – soaking them for the full 12 hours and then leaving them to sprout with rinses for about three days – in which case the resulting sprout will be small with a little curled root, but deliciously sweet. Alternatively you can attempt a home version of the thick-rooted sprouting, remembering that the results will not be so impressive as the bought versions.

Mung bean sprouts grow thickest if they mesh together, or have some outside pressure holding them in place as they grow. A large plastic lidded cup with holes made in the bottom would be ideal as the confined space will push the sprouts together, or you can use a tray sprouter. Put in enough dried beans to fill your sprouter one-third full: you want to create a mass of roots, so start off with much more seed than you would normally. Soak the seed for a full 12 hours, then start sprouting in a fairly dark spot. Place a weight on top of the sprouts and take care not to dislodge them when you rinse, either by using a gentle shower attachment or by dunking the tray and weight into a basin of water and allowing it to drain. Rinse every 12 hours. Once the roots have started to mesh together – around day three – give them a soak for around 20 minutes, then continue with the usual rinsing and draining cycle. Do not rinse for the final 24 hours.

Harvesting and eating

Both types of sprouted mung beans will last for a week or more in the fridge, though they are at their best when fresh. They're very versatile, but classic ways to use them are to add them to stir-fries as the final ingredient, just before serving; to sprinkle them as a garnish for pho, the Vietnamese noodle soup; or to mix them with vegetables and wrap them in spring roll wrappers before deep-frying.

Vietnamese summer rolls

Summer rolls are a lighter, healthier take on spring rolls. Essentially this is a salad packaged up in a rice paper wrap, with dipping sauce to provide the flavouring. You can put whatever you like into the wrap, but mung bean sprouts are more or less essential to provide the crunch. Fat-rooted sprouts and smaller, freshly home-sprouted ones work just as well.

Serves 3

200g (6 ounces) rice vermicelli noodles
6 rice paper wraps
6 large soft, butterhead-type lettuce leaves
large handful of sprouted mung beans
2 handfuls each of roughly chopped fresh mint, coriander (cilantro) and Vietnamese basil
1½ cups shredded roast chicken (prawns or pieces of roast duck or pork work well too)
6 chives

For the dipping sauce

juice of 1 lemon
juice of 1 lime
3 tablespoons fish sauce
2 tablespoons caster (superfine) sugar
2 tablespoons vinegar
50ml (¼ cup) water or fresh coconut juice
3 garlic cloves, finely minced
1 chilli pepper, thinly sliced

1 Soak the noodles in just-boiled water for 10 minutes, then chop them into short clumps.

2 Mix all of the dipping sauce ingredients together and set aside.

3 Lay out all of your wrap ingredients as you will need to work quickly. Dip each wrapper into a large dish of warm water then lay it on a flat surface. Lay a lettuce leaf, sprouts, herbs, chicken, noodles and a single chive at the centre of the top half of each wrap. Fold over to form a semi-circle, then fold each side in. Serve the wraps immediately with the dipping sauce.

Chickpea

3 days

Chickpeas make unusual sprouts with a smooth texture and nutty flavour. If you're making up a salad based on plenty of delicate mixed sprouts, you'll want a few of these more substantial sprouts to provide some extra body. Rich in amino acids and vitamins A and C, chickpeas are also a good source of protein.

Cultivation

As chickpeas are big and robust they are well suited to a sprouting bag. Soak for 12 hours then drain and sprout for three days, rinsing every 12 hours.

Harvesting and eating

Chickpea sprouts are ready to eat from the moment they start to emerge from the large, round seeds. They will store well in the fridge for at least two weeks. While they're great sprinkled on salads they're chunky enough to make a snack in themselves, so leave a bowl of them in the fridge to pick at. They also make a lovely small salad in their own right, lightly dressed with garlic and lemon juice.

Chickpea sprout hummus

This is made just like ordinary hummus, but instead of boiling the chickpeas until they are soft, you sprout them. It has the familiar hummus taste but fresher.

200g (1½ cups) sprouted chickpeas
2 tablespoons lemon juice
2 garlic gloves, crushed
100g (½ cup) tahini
2 tablespoons olive oil
salt

Put all the ingredients in a food processor and blitz to a fine consistency. Serve with flatbreads and salad. You could also top with a sprinkle of sumac or ground coriander and some crudités for dipping.

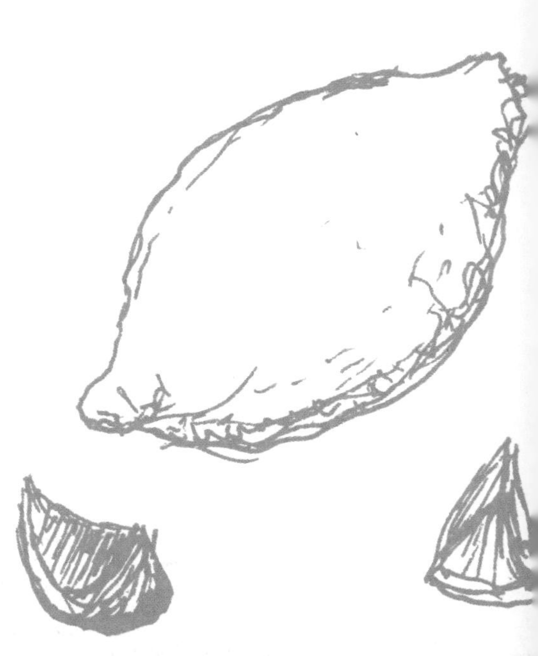

Lentil

3 days

Lentils make delicate and delicious sprouts, particularly nutty in flavour, and high in phosphorus, iron and vitamins.

Cultivation

Soak for 8–12 hours then sprout for three days, rinsing every 12 hours.

Harvesting and eating

Lentil sprouts will store in the fridge for a couple of weeks. Use them to add interest and body (as well as protein) to salads. To make a particularly nutritious dahl, fry onions, garlic, chilli and Indian spices together and add the lentils and a little stock for the last few minutes of cooking.

Pea

3 days

Pea sprouts have the taste and freshness of freshly picked peas, but with more substance and bite. They are also super-nutritious, packing vitamins A, B, C and E, as well as calcium, phosphorus and iron.

Cultivation

Dried peas are large and need soaking for the full 12 hours. Sprout for about three days, rinsing every 12 hours.

Harvesting and eating

Pea sprouts will store well for a couple of weeks in the fridge. Add them to salads and sandwiches or just have a small bowl of them to hand for a snack.

Mark: "I sometimes let peas green up into tiny pea shoots – one of the best things you can add to a sandwich."

Pea and ham sandwich

This is a classic combination of ingredients that works just as well as a sandwich.

salted butter
2 slices of bread
2 lettuce leaves
2 slices of ham
handful of sprouted peas

Butter the bread and lay the lettuce leaves and ham on one slice. Sprinkle with the pea sprouts and top with the other slice.

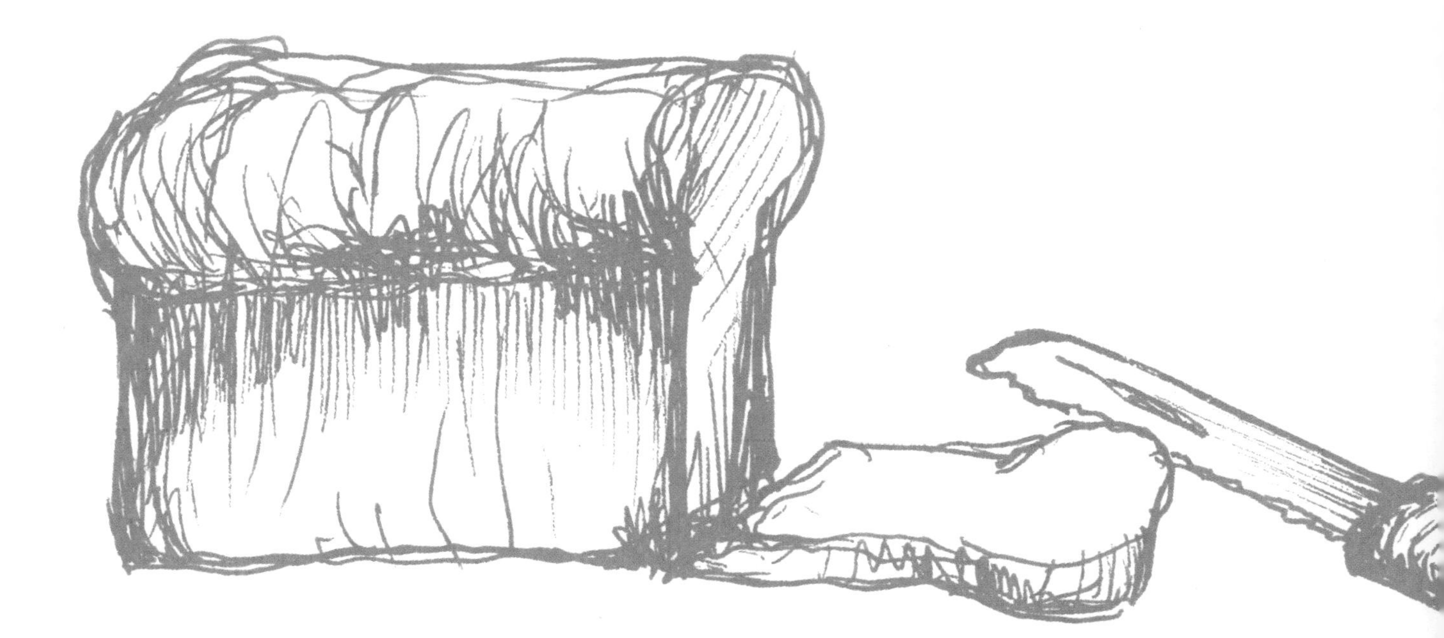

Fenugreek

5 days

Fenugreek seeds are widely used in curried dishes, and when sprouted they have a strong and concentrated curry taste. This can be quite a bitter sprout, but wonderful when mixed with other, milder sprouts, or used delicately in a salad or other dish.

Cultivation

Soak for 8–12 hours then sprout for around three days, rinsing every 12 hours. Fenugreek sprouts are at their sweetest and least bitter when young, so don't grow them too big.

Harvesting and eating

Mix with other, milder sprouts so that the taste is not too overpowering. Sprinkle over curries just before serving to add an extra layer of aroma.

Salad with fenugreek

A simple salad of mild leaves comes alive with a sprinkle of strong-tasting sprouts such as fenugreek.

mixed salad leaves
ball of mozzarella cheese
handful of sprouted fenugreek
about 1 tablespoon vinaigrette

Wash the salad leaves and lay out on a large platter. Break the mozzarella ball into pieces and place over the leaves, then scatter the sprouted seeds and pour on a little vinaigrette.

Lia: "Fenugreek seed is sometimes used as a digestive. I mix a small handful of sprouts into yogurt to make a gently spiced raita to eat alongside Indian food."

Alfalfa

6 days

Bought from a health-food shop, alfalfa is possibly the sprout that puts people off sprouts. It doesn't store well and the texture becomes a little stringy over time, so it's unfortunate that it has become the ubiquitous sprout. However, grown at home and eaten fresh it's crisp and clean in taste; it's one sprout that's very much at its best straight from the sprouter, so if you've tried it from a shop and it's left you dismayed, do give it another chance. Very high in proteins and nutrients, it boasts vitamins A, B, C, D and K as well as a host of trace elements and minerals.

Cultivation

Soak for 8–12 hours then sprout for up to six days, rinsing every 12 hours.

Harvesting and eating

While alfalfa will store for a week, the texture deteriorates on storage so try to use it immediately. It's particularly mild, so do mix it up with other, more flavourful sprouts. It works best as the salad part of a sandwich, adding crunch – try it as a replacement for lettuce.

Micro greens

As well as being the speediest possible route to leafy greens, micro greens are flavour bombshells. Added to salads of larger leaves they impart zing and liveliness, but they can also be used as a salad in themselves or as a flavouring for other foods – they bring a punch of vibrant taste to whatever they are sprinkled on.

Micro greens are just tiny seedlings of plants we usually harvest when they are more fully grown. They are sown into compost and grown in light like any normal seedling, but harvested just a week or so after germination when they've produced their first pair of leaves.

The plants that work best as micro greens are those with intense flavour and/or colour. Coriander (cilantro), basil, fennel, radish and the oriental leaves are all great to try. At micro stage they contain the essence of their fully grown selves, only more concentrated, so you get a burst of flavour, stronger and often cleaner than it would be if you left the plant to grow to maturity. Cutting them so young makes micros feel like a real luxury ingredient – all, of course, would grow on to larger harvests – but that intensity of flavour is the reward, and it comes very quickly.

Growing micro greens is very easy – easier, in fact, than growing plants to greater maturity, since there just isn't time for any pests and problems to get a grip before the leaves are harvested. This makes it a particularly useful technique for usually troublesome crops such as rocket (arugula), which is stalked by flea beetle, or basil, which can rot off. Many of the plants in this section are herbs, so try growing your favourites in this way to see their fresh, lively side. Their youth makes them delicate, so this is not an ingredient for long, slow cooking, but more for adding at the last moment for a little sprinkle of something special.

Mark: "Never use parsnips for micro greens; as seedlings they are poisonous."

SOWING MICRO GREENS

All micro leaves are grown in the same way. Short lengths of guttering are perfect for the job, particularly if you are sowing several different types. They are cheap and widely available, can be cut to whatever length you choose, and contain just the right amount of soil. You could also use seed trays. Choose a fine-textured, peat-free, seed-sowing soil; you are going to harvest the leaves when they are little so you really want a nice smooth surface to make harvesting easier and cleaner. If you use guttering, leave about 5cm (2 inches) at either end or tape up the ends to stop the soil flowing out when you water.

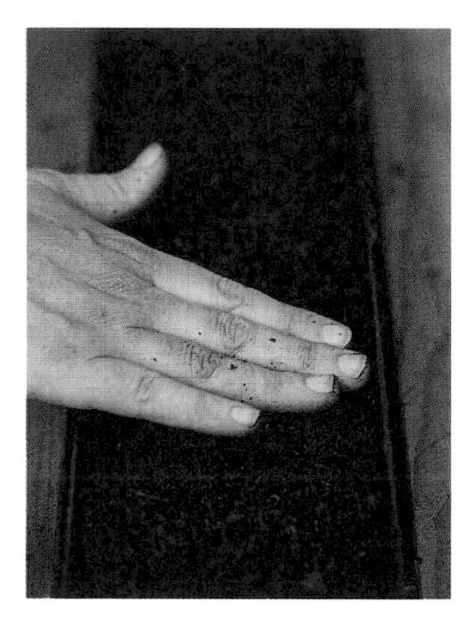

Fill your guttering or seed tray with seed-sowing soil and push down gently with the flat of your hand so that it's well compacted.

Sow a line of seeds down the centre, taking a pinch at a time and distributing them quite thickly but not clumped together.

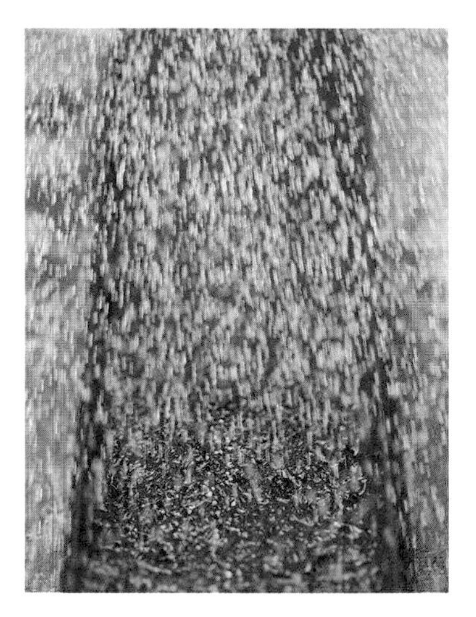

Use a watering can with a fine rose attachment to water them in. Take care not to allow the water to puddle or the seeds may be washed out of line. Water very lightly but often. Although micro greens generally prefer sun, they grow for such a short time that a little shade is not a problem, so place them wherever you have space.

Mark: "To ensure a good variety of micros, I use two lengths of guttering. I sow one with quarters of coriander, rocket, radish and giant red mustard, labelling each quarter. I sow the other guttering in the same way 10 days later. The second sowing should be ready to harvest just as I am coming to the end of using the first, at which point I resow to ensure a successive harvest of micros."

The seeds should be germinated and ready to harvest after about seven days in the summer, or a little longer in winter. Slow germinators such as coriander will be ready in about 10 days in summer.

Pick the micro greens when they're still tiny, pulling them out of the soil roots and all and washing off the soil before eating them; alternatively, cut them just above the soil surface to avoid the need for washing. Ideally, harvest them just before eating; if you have to store them, put them in a plastic bag, mist with water and place in the fridge for up to a day.

Coriander/Cilantro

14 days

There's nothing subtle about coriander, a pungent, leafy herb which is the garnish of choice for bold and full-flavoured Indian and Mexican cuisines. Grown as micro greens, it delivers that punchy, aromatic, unapologetically savoury flavour in even stronger bursts. It has everything we love about coriander, yet with none of the soapiness that often comes with it. When grown for larger leaves, coriander often tries to bolt or run to seed, so it's a great contender for the micro-green treatment as it will never grow large enough to do this.

Cultivation

Coriander can be a frustrating seed to get started, taking a good while longer to germinate than other micro greens – maybe even as long as three weeks. This is partly due to the tough outer coating of the seeds, which prevents water from penetrating. Crush the seeds lightly so that they split in two (between fingers or with a very gentle tap in a mortar and pestle) and then soak overnight to speed up germination and improve success.

Recommended varieties

If you're growing larger leaves (for use in a mixed salad alongside baby leaves, for example) it's worth selecting a variety that resists bolting, such as the imaginatively named 'Slow Bolt'. For micro-green purposes, however, any old coriander will do.

Harvesting and eating

Pull the whole plant out of the soil or snip a bunch at a time. Use to provide a punchy garnish and flash of bright green to finish Indian and Mexican dishes and mix it into guacamole.

Coriander micro greens on baked mackerel

The fresh, strong taste of coriander micro greens works beautifully with this oily fish.

Serves 2

2 whole mackerel, gutted and cleaned
2 bay leaves
dash of olive oil
handful of coriander micro greens

1 Preheat the oven to 190°C/375°F/Gas 5.

2 Lay a bay leaf along the inside of each fish and rub a little olive oil over the skin. Place on a greased baking tray and bake in the oven for about 25 minutes until the skin is browned and the flesh is cooked. Sprinkle generously with the coriander micros and eat immediately with a green salad.

Fennel

7–10 days

Savoury, aniseed fennel – high in vitamin C, potassium and folic acid – is another plant that likes to run to seed at the slightest provocation, and so another that's a perfect candidate for micro-green treatment. If you grow it as a row of tiny, feathery leaves, cut before their prime, you will obviously not get the succulent bulb that forms over a summer of growth, but you will have that aniseed flavour, fast and hassle-free.

Cultivation

Sow as for other micro greens.

Recommended varieties

There's no need to concern yourself too much over varieties, but choose straightforward fennel for silvery-grey leaves or bronze fennel for a little splash of colour with the same strong taste.

Harvesting and eating

Fennel particularly loves white fish, so sprinkle it onto pan-fried or baked fish to finish it. It's also great sprinkled over a fruit salad or with cheese.

Lia: "One of the main drawbacks in sowing micro greens is the cost of the seed: you get through an awful lot more than when you're harvesting more traditionally. I allow a couple of plants to flower in the garden and go to seed, then gather the seed on a dry day in early autumn and store it in an airtight container."

Radish

7–10 days

If you think radishes are quick to grow normally, try them as a micro green – they shoot up almost overnight and are ready in days. They don't need time to develop that hot, peppery flavour: they are punchy from the start but without the harshness that comes with some radishes. You get intensity, clean flavour and no chance of pests or woodiness either. The tiny shoots are rich in ascorbic acid, folic acid, potassium and vitamin B6.

Cultivation

Sow as for other micro greens, but expect to see results quicker than most, certainly within a few days.

Recommended varieties

There's little to choose between varieties for the purposes of micro greens as they are out of the ground before any differences are discernible. However, if you're going to leave them a little longer to eat as roots (they still very much count as quick veg, being ready within a few weeks) see the varieties on page 168.

Harvesting and eating

As the roots of radish in particular have much flavour, pull the entire shoot out of the ground and dunk in water to remove the compost. Sprinkle radish shoots over mild salad ingredients such as butterhead lettuce and cucumber to give them a peppery kick.

MUSTARD
4/2

Rocket/Arugula

7–10 days

The fresher and more youthful your rocket is, the more packed with flavour and good stuff such as vitamins A, C and K it will be, and with perfect leaves to look pretty on your plate. Almost as soon as rocket puts out its leaves, along come flea beetles to drill holes in them. This doesn't happen when you grow them as micro greens, simply because there isn't time for the beetles to find them.

Cultivation

Grow as for other micro greens.

Recommended varieties

Salad rocket, *Eruca vesicaria* subsp. *sativa* (often sold in the UK as *Rucola coltivata*), has relatively mild leaves; wild rocket, *Diplotaxis tenuifolia*, is slightly stronger and more pungent.

Harvesting and eating

Harvest as for other micro greens. Use rocket as a sharpener with eggs or in combination with fresh tomatoes; it's also marvellous in burger buns for a little fresh and peppery salad kick. Rocket salad with Parmesan cheese and olive oil was the dish of the 1990s, but just because it's passé now doesn't mean it isn't still very tasty indeed.

Sow a variety of micro greens in lengths of guttering. You can fit a huge amount of flavour into these compact containers.

Rocket salad with honey and mustard dressing

Serves 2

3 tablespoons olive oil
1 tablespoon white wine vinegar
1 teaspoon honey
1 teaspoon wholegrain mustard
2 handfuls of rocket micro greens, washed and patted dry

Put all of the dressing ingredients into a jar and shake vigorously until they have emulsified. Pour a little over the micro greens and serve as a dainty side salad. Sealed, the dressing will store in the fridge for a few weeks.

Mustard 'Red Giant'

7–10 days

This is one of the oriental greens that's sold as baby leaves, pepping up bags of less punchy salad in the supermarkets. As micro greens the mustard flavour is slightly less strong, but they still provide a hint of wasabi. High in antioxidants, they are also a particularly good source of vitamin K.

Cultivation

Grow as for other micro greens.

Recommended varieties

It's worth experimenting with other hot and spicy oriental mustards as micros, such as 'Golden Streak', 'Red Frills', and 'Green in Snow' – some will be hotter and spicier than others.

Harvesting and eating

Harvest mustard as for other micro greens. Sprinkle or mix into any oriental dishes such as stir-fries, always at the last moment so that they are fresh when served. Dishes such as pad thai, traditionally served with a small portion of crushed peanuts for sprinkling, would benefit from a portion of mustard micros alongside for those who want a more fiery hit.

Sow the prettiest and punchiest of leaves for micro greens, and sow them in abundance.

Oriental greens with egg mayo

Mustard and cress are perhaps the most well-established micro greens, long predating the term, and make a traditional accompaniment to egg mayonnaise sandwiches. 'Giant Red' mustard provides a similar mustard taste without the chilled-sandwich-in-a-plastic-container vibe.

Serves 2

4 large eggs, hard-boiled and shelled
2 tablespoons mayonnaise
salt and pepper
buttered bread
handful of 'Giant Red' or other oriental mustard micro greens

Chop the eggs roughly and mix with the mayonnaise. Add salt and pepper to taste then spread onto buttered bread before topping with a liberal sprinkling of the micro greens.

Red cabbage

7–10 days

A bit of a wild card among the micro greens, red cabbage tastes perfectly good, with the fresh, raw-cabbage taste that you would enjoy in a coleslaw, but is really here for its looks. The tiny shoots are brilliantly, vibrantly purple-red when absolutely fresh and are full of thiamin, riboflavin, folate, calcium, iron and magnesium. They light up a green salad.

Cultivation

Grow as for other micro greens.

Recommended varieties

There is little to choose between the varieties for micro-green purposes. 'Kalibos' has particularly red leaves as a mature cabbage, but as it will be hard to tell them apart from others at such an early stage, sow whatever you can get hold of.

Harvesting and eating

Harvest as for other micro greens. Sprinkle onto green salads or over the top of coleslaw.

Red amaranth

7–10 days

With a fresh flavour and beautiful clear colour, red amaranth is another micro green to grow for contrast with green micro greens and baby leaves. It's a good source of amino acids and vitamins A, B, C and E.

Cultivation

Grow as for other micro greens.

Harvesting and eating

Harvest as for other micro greens. Red amaranth adds a subtle depth and variety to other, stronger flavours, complementing and enhancing micro greens such as coriander. The red is great in visual combination with bright greens. Sprinkle over salads to turn them into a visual feast: they look particularly beautiful sprinkled over edible flowers.

Lia: "Larger amaranth leaves are a common ingredient in Indian and Chinese dishes, so the micro greens sit easily with these cuisines. I scatter them on top of dhals and stir-fries for a little extra texture and colour."

Red Amaranth

Basil

7–10 days

Even the least herb-tolerant toddler loves pasta and pesto, and it's the easy herb for adults too; few can resist an evocatively Mediterranean plateful of sun-warmed tomatoes and cool mozzarella scattered with torn basil. The micro-green version contains all that savoury Mediterranean flavour but in greater intensity (the distinctive flavour is the result of the essential oil, which has anti-inflammatory and anti-bacterial properties).

Many people find basil a tricky herb to grow – its succulent stems rot off at the first whiff of overwatering and it's prone to all sorts of diva-like behaviour when grown to full size. Not so as a micro green, though, as it doesn't get time to misbehave. Having said that, you can grow basil a little larger than some of the other leaves, since nothing is lost of that aromatic flavour as the leaves expand.

Cultivation

Grow as for other micro greens.

Recommended varieties

There are a number of different basils you can try as micro greens. Sweet basil is the most widely available and often considered the tastiest and most suitable for Mediterranean dishes; purple basil is similar in taste, but with pretty purple colouring so that it looks particularly good against other salad leaves. Thai basil has a more aniseed taste and is better suited to Eastern dishes than Mediterranean.

Harvesting and eating

Harvest as for other micro greens and scatter onto salads, particularly tomato salads, since basil has a great affinity with tomato. The purple basil looks (and tastes) particularly good in goat's cheese or mozzarella salads.

Edible flowers

Growing edible flowers feels a little like cheating, since flowers are usually just a stage in the cropping process. You want your vegetable plants to flower, sure, but you're most interested in that for the pollination that results in the food for your table. However, in the case of the plants in this chapter those flowers are all we want, for their unusual tastes and textures and their visual appeal.

Edible flowers are quick and delicious and look great both in the vegetable garden and on the plate. They are a particular boon for salad-lovers, and most of those mentioned in this chapter are at their best treated simply, just picked from the garden and sprinkled onto mixed leaves. Here they add jewel-like beauty to what can be the duller part of the meal, studding straight green leaves with flashes of purple, orange, yellow and blue. Many straddle the sweet and savoury worlds, and are particularly pretty crystallized and sprinkled onto desserts.

They're not all about looks, though. Each has a taste that adds to the layers of interest on the plate, be it peppery, cucumber-like or fresh and sweet. There's very little bulk to these beauties, but they lift other dishes immeasurably and set them out as unquestionably home grown.

Nasturtium

50 days

Even if nasturtiums didn't produce flowers they would be grown for their leaves – circles of bright, fresh green, draping themselves elegantly along flower beds and smothering weeds as they do so – but their flowers are very special, produced in brightest oranges, egg-yolk yellows and deep burgundy reds. And they're edible: you can just pop them into your mouth as you wander round the garden. The flavour is complex, with a little spice, a little honey, and a little pepper chasing along after. Both flowers and leaves have this peppery taste, and the young leaves can be used in salads or ground up with toasted walnuts, garlic, olive oil and a hard cheese such as Parmesan to make a nasturtium pesto.

Nasturtiums are one of the flowers that looks most at home on a vegetable plot, so plant plenty and let them range about. You'll certainly find plenty of uses to put them to once you've discovered how delicious they are.

Cultivation

Nasturtiums are tender annuals, so sow them indoors or in a greenhouse a couple of weeks before the last frost. Pot them on into small pots a few weeks after germination and plant out somewhere sunny once all danger of frost has passed. You can also sow them direct if you wait until later in the spring; just push the seeds into the ground, leaving around 20cm (8 inches) between them. Slugs may be a problem in the early days, so give the young plants some protection.

While the stems and leaves of the plants are the tenderest of things (their collapse at the first whiff of autumn frost is a sign each year that winter is truly on its way), the seeds will often survive year on year so with a bit of luck you should only need to plant once.

Recommended varieties

***Tropaeolum majus* Alaska Series** Colourful yellow, orange and red flowers against white-variegated leaves.

***T. majus* 'Empress of India'** Ruby-red flowers set off against darkest green leaves.

***T. majus* 'Milkmaid'** A pale primrose flower against dark green leaves.

***T. majus* 'Tip Top Mahogany'** Dark flowers and particularly peppery leaves.

Harvesting and eating

You can start to harvest the leaves as soon as they are large enough and the flowers as soon as they appear. Pick and eat them as a snack while you garden, or gather handfuls and scatter them on salads. They are also wonderful dipped in tempura batter and deep-fried.

Lia: “After picking nasturtium flowers I float them, faces upturned, in a bowl of water until I’m ready to use them. It keeps them fresh and even plumps them up a little.”

Salad with nasturtium flowers

Serves 2 as a side dish

2 handfuls of mixed leaves
1 tablespoon vinaigrette
10–12 nasturtium flowers

Make up a shallow bowl of mixed salad, dress with the vinaigrette, then scatter the surface with freshly picked flowers. Eat immediately.

Daylily

Perennial, flowering throughout summer

Much loved by ornamental gardeners for their large, showy flowers, daylilies are undeniably a border plant. This makes them a little different from the other edible flowers, which are often residents of the vegetable garden anyway and so allow us to make the mental leap from plot to plate with ease. It may seem a little strange at first to pluck a flower from your border and eat it, but daylily flowers really are delicious.

In fact all parts of the plant are edible, but it's the flowers that are the most tasty part, with a satisfying crunch and a flavour somewhere between sweet lettuce, radish and green bean. They are produced in abundance, up to 20 flowers on each flower spike for week after week every summer, so on an established plant even the most enthusiastic harvesting won't make too much of an impact on the show.

Cultivation

Daylilies are perennials, so all you have to do is plant them and get them established and you will have a source of flowers every year. Plant in autumn or spring in rich soil in dappled shade or full sun. Daylilies like moisture, so you may need to water them during droughts, but they are really very tolerant once they have their roots down.

Recommended varieties

Hemerocallis citrina This should really be called a night lily as its yellow star-shaped flowers open late afternoon and bloom into the evening, emitting a delicious scent as they do so.

H. fulva A vigorous, clumping kind that will quickly spread. Use this as a groundcover plant if you have lots of space to fill.

***H.* 'Pink Damask'** Salmon-pink flowers in mid-summer.

***H.* 'Red Rum'** One of the tastiest of the red varieties, and compact too, so good for planting in pots.

Harvesting and eating

The genus name *Hemerocallis* derives from the Greek *hemera kallos*, or 'day beauty', alluding to the fact that each flower blooms for just one day then fades. The best time to pick the flowers therefore is towards the end of the day, just before they start to wilt, so you can enjoy both looks and taste. The buds are also good, with a taste somewhere close to a green bean, but they sweeten up as they open and bloom.

First try them raw as a snack, or sauté them in butter with a little salt and pepper. You can also dip the flowers in tempura batter and deep-fry them. A small minority of people may have a bad reaction to eating daylilies.

Mark: "Generally I have found the yellow-flowered varieties to be the sweetest – a little more bitterness creeps in with the red-flowered varieties."

Viola tricolor

60 days

Commonly known as heartsease for its medicinal properties, *Viola tricolor* has long been used as an edible flower, sprinkled on salads or desserts. It makes a dainty plant in the garden, with pretty little violet and yellow miniature pansy flowers possessing a delicate, perfumed flavour.

Cultivation

Sow seeds in early to mid-spring under cover, then plant out at a spacing of around 20cm (8 inches) once the weather has started to warm, protecting young plants from slugs. You will be able to harvest flowers from this sowing from summer onwards. Violas are hardy plants and can also be grown for winter flowers in mild climates. Sow your winter plants in late summer, planting out in autumn. A cloche placed over the plants will protect the delicate flowers from the worst of the winter weather.

Harvesting and eating

Pick the flowers when they have just opened and float them in cold water to swell up slightly before scattering them onto salads. To crystallize the flowers, use a fine paintbrush to coat them with egg white and then scatter caster sugar over them. Leave them to dry for two hours. Use them within a week, on top of cakes or other desserts.

Courgette/Zucchini

35 days

While courgettes fruit quickly enough to appear elsewhere in this book, the flowers are even quicker, and are such a treat. They're one of those ingredients you only get the chance to eat if you grow your own food, since they're just too delicate to last in the shops.

Cultivation

Sow courgettes indoors in mid-spring; they are tender plants and cannot take the slightest hint of frost. Pot them on as they grow, and when the weather warms and there's no chance of further frosts, start acclimatizing the plants to the outdoor life by leaving them out for increasing periods of time during the day. Eventually they'll be able to stay outdoors overnight and then they are ready to plant out. Protect them from slugs while they are young and soft by planting them in a copper ring. Water well throughout the growing season, particularly during dry spells.

Recommended varieties

'Bianca' This variety produces unspectacular courgettes but a great number of flowers, making it worth growing just for those. In the USA, this variety goes by the name 'Bianco di Trieste'.

Harvesting and eating

Harvest early in the day when the flowers are young and fresh. Wash carefully and remove the central stamens. You can simply fry the flowers in butter, season them and eat immediately as a starter or side dish. They have a delicate courgette taste and a lovely texture.

Mark: "Learn to spot the difference between the male and female flowers. Females have a small bulge behind the flower that will turn into a courgette. Males just have a straight stem and these will never be courgettes, so pick them with abandon."

Stuffed tempura courgette flowers

Stuffed and deep-fried in batter, courgette flowers are delicious. You can use cream cheese mixed with herbs for the stuffing, or, as here, cold risotto.

Serves 4

12 courgette flowers
2 cups cold risotto
sunflower oil for deep-frying
85g (⅔ cup) plain (all-purpose) flour
½ teaspoon salt
200ml (⅞ cup) ice-cold water
Coarse sea salt, to serve

1 Remove the centres from the flowers and use a teaspoon to stuff each about two-thirds full with risotto.

2 When they are all prepared, start warming the oil in a deep-sided pan.

3 Meanwhile, make the batter, which should never be left to stand. Sift the flour and salt into a bowl and briefly whisk in the ice-cold water. Drop in a few ice cubes to keep it cool. Don't worry if there are lumps of flour left in the mix as this adds to the crispiness.

4 Once the oil is hot (a piece of potato dropped in during heating will rise to the top when the temperature is right for frying), twist the tips of the flowers together and dip into the batter before lowering carefully into the oil. Fry for a few minutes, turning the flowers so that they are evenly brown on all sides. Lift out, drain on kitchen paper and serve immediately, sprinkled with coarse sea salt.

Borage

50–60 days

Borage produces star-shaped white or sapphire-blue flowers with a fresh, cucumber taste, perfect for adding to salads and to summery drinks. Even without its edible qualities it would be a great plant for the vegetable garden, as its flowers are magnets for bees. Plant it near to cropping plants that will benefit from their pollination – tomatoes, strawberries, beans, courgettes and squash. It's even said to improve the taste of tomatoes when grown nearby.

Cultivation

Borage is a hardy annual that grows quickly to a slightly ungainly 90cm (3 feet). Sow seeds under cover in early spring and plant out in early summer, in a sunny spot with well-drained soil. You may need to give the plants some support as they can get a little tall for themselves and flop over, but full sun will help them to develop a sturdy habit. After the first year, borage should self-sow freely, so you are unlikely to have to sow twice. Just pull out and compost the old plants as they die off.

Harvesting and eating

Snip off the young flowers and scatter them on salads, onto the surface of cold soups (they are a particularly good addition to cucumber soups) and into summery cocktails. To make crystallized flowers for decorating cakes and puddings, paint them with egg white, dust with caster sugar and leave them to dry.

Lia: “As well as making use of the flowers I also add young borage leaves to salads: they have the same cucumber flavour. Older leaves can be used as a spinach substitute.”

Agua de Valencia with borage ice cubes

Borage is delicious added to fruity, summery drinks such as this Spanish cocktail which celebrates Valencia's greatest export: oranges. Always use freshly squeezed juice.

borage flowers
water
200ml (⅞ cup) freshly squeezed orange juice
75cl bottle cava
60ml (¼ cup) gin
60ml (¼ cup) vodka
pinch of sugar
slices of orange

1 Fill ice cube trays about one-third full of water and drop a borage flower into each. Freeze. Top up with water to cover the flower and freeze again. Freezing in two stages ensures that the flower is at the centre of the cube, rather than floating on the top.

2 To make the Agua de Valencia, pour all of the ingredients except the orange slices into a pitcher, mix gently, then place in the fridge until ice cold. Serve with the orange slices and a borage ice cube.

Chives

Perennial, flowering early summer

Chives are a well-loved herb, the green blades scissored into neat little rounds on top of scrambled egg and mixed into mashed potato. But the pale purple flowers, produced from late spring, are equally tasty, and pretty too when sprinkled on top of dishes. They attract bees, so are great for planting near vegetables and fruit that need pollinating early in the year.

Cultivation

Chives are one of the few herbs that grow well in shade, although they will produce more flowers in a sunnier spot. They like fairly moist ground, so you may need to water during dry spells. As they are perennial you should only need to plant once.

Harvesting and eating

Pick chive flowers just before eating, so that they are as fresh as possible. They have the same onion taste as the leaves do, but milder, so they work well with raw, savoury dishes as well as looking beautiful on salads – break up the flowerheads and sprinkle them all over.

Chive flowers are a fleeting treat, but you can preserve them by making them into a herb butter. Mix the broken flowers into butter then roll the butter into a cylinder and wrap in cling film before placing in the freezer. You can later slice off and thaw exactly the amount of butter you need. Spread it onto good bread or melt onto risottos or pasta dishes as a finishing touch.

Pot marigold

40–50 days

Pot marigolds (*Calendula officinalis*) produce cheerful orange flowers that light up the vegetable garden. Although they most probably originated in southern Europe, they are among those plants that have been cultivated for so long that their origins in the wild are unclear. The flowers have a great many medicinal uses and of course are also good to eat, with a peppery taste and a colour that looks beautiful sprinkled over food, particularly salads.

Cultivation

Calendula officinalis is a hardy annual so in theory it can be sown where it is to flower, in early to mid-spring. However, you will get quicker growth and stronger, earlier-flowering plants if you start them off under cover and plant them outdoors once the weather has started to warm and any chance of hard frosts has passed. Plant them in full sun, in rich but well-drained soil, and they will quickly put on growth and start producing flowers. You may find they self-seed the following year, but again for the earliest possible flowers you may want to collect some seeds on a dry day in autumn and sow afresh in spring.

Harvesting and eating

Pick flowers as you need them and use straight away as a sunny garnish on soups and salads. They will not store.

Calendula and monkfish paella

This is adapted from the Paella de rape con azafrán recipe in *Moro: The Cookbook* by Sam and Sam Clark (Ebury Press, 2003). We have used marigolds here instead of saffron to colour and decorate the dish and impart a mild pepper flavour.

Serves 4

good handful of calendula petals
120ml (½ cup) olive oil
400g (14 oz) monkfish fillets, trimmed and cut into bite-sized pieces
2 large onions, finely chopped
2 green peppers, deseeded and finely chopped
6 garlic cloves, crushed
½ teaspoon fennel seeds
250g (1 heaped cup) paella rice
80ml (⅓ cup) fino sherry
800ml (3⅓ cups) fish stock
small bunch of flat-leaved parsley, roughly chopped
½ teaspoon sweet smoked paprika
salt and pepper

1 Put half the calendula petals into a mortar and pestle with 2 tablespoons of olive oil and grind to a fine paste.

2 In a large frying pan, sauté the monkfish in 2 tablespoons of the oil until just undercooked. Transfer the monkfish and juices to a bowl and set aside.

3 Add the remaining oil to the same pan and fry the onions and peppers until soft, then add the garlic and fennel seeds and cook for about 3 minutes.

4 Add the rice and stir in for 1 minute before adding the sherry then the stock, monkfish, half the parsley, the paprika, the calendula paste and seasoning to taste. Bring to the boil then reduce the heat and leave to simmer for about 15 minutes, uncovered. Turn off the heat, cover with foil and leave to rest for 5 minutes. Decorate with the remaining parsley and calendula petals and serve.

Lavender

Perennial, flowering early summer

Added with a light touch, lavender can be a magical ingredient, sprinkling dishes with the essence of the summer garden. The leaves and flowers are packed with volatile oils that release their scent in warm weather or on being bruised; used with too much abandon they can produce a soapy taste and their strong camphor notes can be unpleasant, so take care to impart just a delicate ghost of lavender to dishes for a really special effect. It is a beautiful plant, grown almost as much for its mound of silvery leaves that look good throughout the year as for its purple spikes of flowers, beloved by bees.

Cultivation

Lavender is a short-lived Mediterranean sub-shrub that hails from dusty, dry and hot conditions. Transplanted to cooler, wetter soils it will sulk and eventually die. It particularly hates having wet roots in winter, so it does beautifully in a container of potting soil mixed with gravel or grit for really sharp drainage. Full sun, it goes without saying, is essential. Silver-leaved shrubs such as lavender will not put up with hard pruning, so the trick is to prune a little every year just after flowering, removing the old flowers and about 5cm (2 inches) of growth, but no more.

Recommended varieties

Lavandula angustifolia Has less strong camphor notes than some lavenders, and a more fragrant, sweeter taste.

***L. ×intermedia* 'Provence'** The best-flavoured lavender for cooking.

Harvesting and eating

Pick the flowers just before they open for the best flavour. Make a lavender sugar by mixing a few flowerheads into a jar of sugar and leaving them to infuse for a few days. You can then use the sugar to make lavender-infused cupcakes and shortbread, or a delicate honey and lavender ice cream. In Provence, lavender is used as a savoury herb much like rosemary. Try it as a rosemary replacement with slow-roast lamb.

Cut-and-come-again salad leaves

One thing you should be able to produce with aplomb in a very short time indeed is a salad, and a colourful salad packed with a whole heap of varying textures and flavours at that. Salad leaves are at their best when little and tender, plucked the day they are to be eaten.

The best possible way to get salad leaves that fit this impressive bill is to use the cut-and-come-again technique. A great range of salad leaves can be grown using this method, giving you the option of a single-variety salad or a salad bag created to your own blend. Typically, lettuces such as 'Reine des Glaces' will be there to provide the bulk of the salad, mild, sweet and soft green, while leaves such as 'Giant Red' mustard punctuate it with hot, peppery notes. The end result is yours to choose, much cheaper to produce, undoubtedly fresher, and vastly more interesting and varied than what's available in the supermarket.

What you won't get when using the cut-and-come-again technique are big, whole, headed lettuces. Some people love to let a lettuce heart up and enjoy the pale, sweet leaves in the middle, but there are drawbacks with producing whole lettuces. A row will often mature all at once, whereupon they will all need to be eaten pretty quickly or they will bolt (flower and then set seed) and the leaves may become bitter. Consequently, it's easy to become overwhelmed by a glut, quickly followed by an absence. This should never happen with cut-and-come-again leaves, where you harvest a little at a time, and harvesting only encourages another flush of leaves.

The range of plants that can be grown as a cut-and-come-again crop is wide, and with a little planning you should be able to crop small, succulent, interesting leaves for much of the year.

For beautiful, delicious and quick salads, sow cut-and-come-again leaves such as green 'Oak Leaf' and red 'Deer's Tongue'.

CULTIVATION

Start with a seedbed raked to a fine tilth. It pays to spend a little time getting the soil just right, since these are little seeds that grow into little plants and they will struggle to do so if they germinate between big lumps of earth. You can also sow cut-and-come-again seeds into containers of potting soil; they look good and you can then keep them near at hand, by the kitchen door. The best containers for the job are shallow and wide, as none of the plants will need a large root run and you'll be able to fit plenty in. Wooden vegetable boxes or old wine boxes are perfect as they aren't so shallow that the potting soil will dry out quickly.

Rather than sowing in a straight line, all cut-and-come-again seeds should be 'broadcast sown'. This means scattering them over a fairly wide area. If you're doing this in a container of potting soil, just scatter them over the whole surface. If you are sowing in the ground, use the back of a rake to draw a wide, shallow drill (furrow) through the soil and sow across that. The seeds will be much closer together than you would ever sow any plant destined to be grown to full maturity, but that's no problem as these will never need the space in which to spread themselves out.

In summer, these plants will grow quickly and be quite happy out of doors but in autumn and winter, growth will be much slower. Those that produce winter leaves are perfectly able to withstand the cold, but they will get battered by rain and frost and the leaves will toughen up, so give them a little protection if you want the sweetest, loveliest leaves. You will get the best results if you can grow them in a greenhouse, polytunnel or cold frame, but if the open ground is your only option you will get improved results by covering leaves with a cloche or even just a piece of horticultural fleece (floating row cover, available in the United States as Reemay).

Slugs, being lovers of the soft, young and leafy, are going to be a problem for all these plants. You will need to protect seedlings or they will be grazed to the ground the moment they put their heads above the parapet. Tiny microscopic nematodes that are watered onto the soil and attack slugs from the inside are useful in keeping the general population down, although it has to be a pretty bad way for a slug to go.

Some form of direct protection will be necessary too for the ones that get past the nematodes. Organic slug pellets based on ferric phosphate are both effective and harmless to wildlife other than slugs. Planting in containers will make it easier to keep slugs away (they struggle to cross copper so try the thin copper bands designed to be stuck to the edges of pots). Plants will also fare better once they are beyond the seedling stage, so you could start off plants indoors and plant them out later, when they'll no longer be the most delicate and slug-tempting thing around.

Sow seed across the surface of your pot, then cover with a thin layer of compost and water with a fine rose attachment.

HARVESTING

Harvest cut-and-come-again leaves when they're 7.5–15cm (3–6 inches) high. Grab a clump of leaves with one hand and a knife with the other and cut as if you were giving the tray a fairly severe haircut, leaving uncut bits. You want the plants to be able to sprout again, so the place to cut is just above the tiny new leaves at the base of the plant; if you cut off this growing point you'll remove the plant's ability to grow more leaves. Done right, though, the plant should regroup and send out another few leaves which will be ready within a couple of weeks. Once again, you cut them just above the growing point and let them regrow.

They will put up with this treatment for three or four cuts before the succulence starts to give way to a little toughness. At this point you need another tray already up to speed and ready for cutting. Keeping a succession of cut-and-come-again leaves in the wings is really the only tricky thing about growing them, and in fact the 'tricky' bit is simply remembering to resow. You will need a system: sow a replacement tray or row just before you make each first cut of leaves, and you should never be left wanting.

Many cut-and-come-again crops naturally grow as a small, loose rosette, and another way of harvesting is to pull individual leaves away from the outside of the rosette, leaving the centre of the plant intact. This is a great way of making up a small mixed salad of different leaves.

CONDITIONING

While most vegetables are at their best the moment they're picked, salad leaves are an exception. No matter how well you care for your salad plants, chances are that they'll be a little floppy once cut – particularly if harvested in the heat of the day – and they'll benefit from conditioning, a little time spent in water to crisp up. Cut them straight into a plastic bag and seal it, then tip them out into a bowl of ice-cold water as soon as you get them into the kitchen. Their cells will slowly but surely fill up with water. Let them float there for a good hour or two before draining and spinning them dry. They will have all their crisp and crunch potential fully restored, and will store much better – up to a week – in a plastic bag in the bottom of the fridge.

HARVESTING AND EATING

The flavour will develop as the leaves grow, so try at different stages to see when it suits you. Just pinch a leaf off and have a nibble. Use a knife to cut enough to provide you with salad for a couple of days, then condition and eat, or store for a few days in a sealed plastic bag in the fridge.

Cut-and-come-again leaves can be grown close together and cropped several times, making for a wonderfully productive use of space.

Oriental mustards

21 days

These are the perfect plants to be given the cut-and-come again treatment: beautiful, crisp and flavourful. Grown into large leaves they are strong and fiery and really must be cooked in order to mellow them down, but when cut young they add just the perfect touch of peppery heat to milder salads. They are among the best-looking leaves, and will be essential if you want good-looking salads. And you should: there's nothing more likely to tempt you to eat your greens than a fine-looking salad of myriad textures and colours.

Cultivation

While oriental mustards are both hardy and heat-tolerant and can thus be made to grow year round, they do particularly well in cool conditions and winter is the time they'll serve you best: when all else is dead and brown, they are verdant and bright and ready for picking. They do vary in hardiness and their ability to withstand winter conditions, though – grow 'Green in Snow' for a surefire winter green.

Spring to midsummer is the oriental mustards' Achilles' heel. When differences between night and day temperatures are at their most extreme they are prone to bolting, after which the leaves become less tender and more bitter. They are also sensitive to day length and try to flower as the days lengthen towards midsummer, so you are likely to get better-quality leaves if you make sowings from midsummer onwards. It makes sense to concentrate on growing them as a major feature of late summer salads.

These leaves will do well in well-drained soil in an open, sunny position, but sow them in shade in summer to discourage bolting. They are pretty hardy plants and so can be wonderfully productive throughout winter. They don't put on a great deal of growth once cold weather hits, though, except in mild spells, so the trick is to make extra sowings in late summer that you'll then be able to use over winter. You'll be able to make four or five cuts of oriental mustards before they need replacing.

The oriental mustards are not just tasty, they are among the most beautiful and varied of the baby leaves you can grow.

Mark: “Flea beetles jump (like fleas) when disturbed, and you can reduce the population by running a piece of card smeared in treacle along the tops of the leaves. As they jump, they stick.”

While the leaves are hardy, they will get battered around by winter weather. Grow them under cover for the loveliest leaves. Even just a cloche or a length of horticultural fleece will make a huge difference to their quality, and to your appetite for them.

Being members of the brassica or cabbage family, they are besieged by flea beetles: tiny, jumping creatures that leave minute holes all over the leaves. They are perfectly edible after an attack, but not especially appealing. Very occasionally the damage can stunt growth to an extent that plants die, but this will happen only if the attack is severe and the plants small; they usually outgrow the damage eventually. Cutting plants when they are so young helps to a certain extent, but there's usually plenty of time in a cut-and-come-again plant's life for significant damage. Try covering your row of plants in horticultural fleece, pinned into the ground on either side. It can be worth sowing an early spring crop just in order to miss the main flea beetle season, which usually hits around midsummer. You will also need to protect these plants against slugs.

Recommended varieties

Mustard 'Giant Red' If looks are important to you this is a good mustard to have in your armoury, with its dark red-purple puckered leaves laced with lime-green veins. The under-

Lia: "If any of these spicy oriental mustards bolt in summer, make use of the flowers – they're edible and savoury, making a good garnish for a Bloody Mary or a soup. Later I also collect up the seed and resow it for peppery micro greens."

Mustards 'Giant Red' (left) and 'Green in Snow' (centre and right) are two of the many oriental leaves that are worth trying as baby leaves.

surface of the leaves is also pale green. It's a particularly dark leaf that's beautiful against pale lettuces. Among the strongest of the mustards as well as being a beauty, it is well deserving of a place in the salad bowl. It's a little slower to bolt than some of the others, so you may be able to get a decent summertime crop. The leaves have a nicely peppery taste and become very strong once they get larger, when you can use them in stir-fries and soups.

Mustard 'Green in Snow' As its name rather poetically suggests, this is one of the best salad plants you can grow in winter. It grows faster than its relatives, its strong, substantial, fresh green leaves coping with all that winter can bring. It will of course withstand being snowed upon, though you will get much nicer leaves if you give it some protection.

Mustard 'Green Wave' A mid-green mustard with deeply frilled edges. Like other mustards it is tolerant of cold and will overwinter well, but it also copes with heat, making it good for summer leaves.

Mustard 'Red Frills' Strongly serrated leaves make this mustard look a little like rocket, but with a purple tinge to it. It is a lovely-looking leaf, and not too strongly peppery either. Use it to brighten up and spice up plainer and paler leaves.

Mizuna Introduced to Japan from China, mizuna was originally grown in the vegetable gardens around Kyoto. As such it has a special place in Japanese culture and cuisine. It has a tall, thin, white leaf stalk and feathery green leaves. Spiky in shape and mustard in flavour, it appears plentifully in supermarket bags of mixed salad leaves – but you'll never find it on its own unless you grow it, when you can really appreciate its fresh, peppery taste. It's good at withstanding winter weather.

Mibuna Named after the town of Mibu, mibuna has a slightly stronger taste than mizuna and grows less vigorously. It is slower to bolt than is mizuna when the temperatures rise in early summer. It's not quite as hardy over winter as some of these leaves are, but it does well under cover. Mibuna has smooth leaves that are tall, slender and mid-green with a white mid-rib. It looks beautiful cut and mixed with the more intricate leaves of mizuna.

From left to right, mustard 'Ruby Red Streaks', mustard 'Red Frills' and mizuna, grown as cut-and-come-again leaves.

Komatsuna This is another leafy vegetable of the brassica family, but much milder than the others. While it's commonly referred to as Japanese mustard spinach, it leans more towards spinach than mustard in intensity but is packed with even more vitamin C, iron, calcium and other nutrients than spinach. The leaves are a glossy green, crunchy and mild but with a slight bitterness that works beautifully with seafood, so do try steaming or lightly sautéing the larger leaves and serving them with fish. The leaves don't collapse like spinach does on cooking so they retain a delicious crunch. Komatsuna is also great raw when the leaves are still small and makes a good base salad leaf against which the spicier leaves can shine. It's particularly good for use over winter.

Harvest just what you need with a sharp knife or a pair of scissors. Cut high enough and the leaves will grow back. You can do this several times before the plants are exhausted.

Harvesting and eating

Start trying the leaves when they are fairly small, as the flavour is hot and peppery and becomes more so with age. Each of these is differently hot too, with 'Green in Snow' being at the milder end, and 'Giant Red' one of the hottest, so try nibbles of each of them as they grow, and identify the point at which you like them best of all.

After picking them, condition the leaves in a big bowl of cold water for an hour or two and then eat or store in the fridge for a few days. Combine them with a big pile of mild lettuce leaves in a salad. If you find the leaves have become a little too strong for your taste to be enjoyed raw, cooking moderates their heat; chop them into a stir-fry or shred them into winter soups.

Oriental leaves make a strong, textured and attractive salad.

Mark: "Try the mibuna variety 'Green Spray' – it's lighter in colour than regular mibuna and has a fresher, milder flavour."

Mibuna is tasty and hardy, with a simpler leaf shape than many of the oriental brassicas.

Kai lan

60 days

An unusual and pretty plant that's worth seeking out, kai lan has succulent blue-green stems, with a taste somewhat resembling broccoli, kale and asparagus: Chinese broccoli and Chinese kale are among its common names. Nutty and sweet, it has a slight mustard edge. It's not strictly a cut-and-come-again crop in the same way as the other leaves mentioned in this section – it grows as a small broccoli-like bush which you harvest by grazing the young leaves, stems, flowerbuds and flowers. The result is a true delicacy. Unlike the other plants in this chapter it's perhaps not at its best when eaten raw, but should be lightly steamed.

Cultivation

While kai lan is a hardy perennial, it's most often grown as an annual. Sow seeds into small pots in early spring under cover and plant out the seedlings at a spacing of about 40cm (16 inches), ideally in sun, though it will tolerate some shade. You can also sow direct into the soil in early to mid-spring, thinning out the resulting seedlings to 40cm (16 inches) later on. Each plant will grow into a fleshy bush, up to 1m (3¼ feet) tall if you let it, but more likely about half that once you develop a taste for it. Prune by almost constant harvesting.

Kai lan is bothered by the same pests that afflict members of the cabbage family, such as cabbage white caterpillars, but they don't seem to inflict the same levels of damage as they do on other brassicas. There may be a little cosmetic damage, but you should be able to find unaffected leaves. It's worth checking the plant occasionally throughout the summer, looking for eggs laid on the undersides of the leaves and squishing them if you find them, since it's the caterpillars that do the damage once they hatch. Alternatively, you can cover the entire plant in fine-meshed netting, which will completely prevent the butterflies from laying their eggs on the plants, but it looks a bit ugly and isn't really necessary.

Clubroot – a practically untreatable disease of brassicas that persists in the soil for years – is a bigger problem, so if you plan on resowing from scratch each spring, make sure you practise rotation, planting in a different part of the vegetable patch every year to prevent a build-up of the disease in the soil. Strong-growing small plants are far better able to resist club root than tiny seedlings, so if you know you have club root in your soil, nurture plants under cover and plant out later, rather than sowing direct.

If you wish to do so, there's no reason not to leave kai lan in the ground at the end of the season. It will flower towards the end of summer (eat the flowers too) and should prove perfectly hardy over winter. Cut it back hard the following spring and you will have those delicious shoots with half the work.

Harvesting and eating

Every part of kai lan is edible, leaves, stems, and flowers too when they appear resembling tiny broccoli stems towards the end of summer. However, the leaves and the tender ends of the stems are the best bit, picked young and soft. They can be eaten raw, but are better when lightly cooked: steam them, sauté for a few minutes in butter or use in stir-fries. Once steamed, the stems are wonderful as the 'soldiers' dipped into a soft-boiled egg, or covered with hollandaise sauce. Anything you would do with a spear of asparagus you can do just as well with a stem of kai lan.

Spinach

30 days

Of all the mild, leafy vegetables you can grow, spinach is the finest: sweet yet sharp in taste and smooth in texture. It is perfect grown small as a raw salad vegetable, and it's rounded, smooth, dark green leaves will be the basis for many a salad. These are the mild-natured leaves against which your more extravagant and adventurous leaves can show themselves off, but they are a gourmet mild salad leaf and certainly good enough to make a salad in their own right. They can also be cooked, though you will need great armfuls of them as they collapse to a fraction of their raw size as soon as they hit heat. No matter, the taste on cooking is arguably even better, and the leaves are still exceptionally good for you, containing high levels of vitamin C and iron. Choose the right varieties and you'll be picking spinach even through winter in mild climates and during the notorious 'hungry gap' of early to mid-spring in cooler zones of the USA.

Cultivation

Sow direct into the ground once a month from mid-spring until early autumn, covering early and late crops with a cloche. Sow fairly densely across a wide seed drill or, for slightly larger leaves for cooking, sow along a line and thin out to about 20cm (8 inches) apart. Spinach plants can often bolt (run to seed) in the heat of the summer, after which the leaves take on a bitter taste; growing in light shade and keeping up with watering during dry spells will help. Towards autumn, increase sowings so that you leave yourself plenty of swathes of spinach to cut once cool weather slows growth. Cover these later sowings in late autumn for the best-quality leaves. Protect from slugs, especially just after sowing when the seedlings are young and tender.

Recommended varieties

'Dominant' Hardy for autumn and winter leaves. Sow plenty in early autumn for eating over winter.

'Matador' One of the best varieties for coping with summer heat. Resistant to bolting.

'Space' A good, slow-to-bolt variety with slightly savoyed leaves. Quick-growing for baby leaves and does well sown in spring, summer or autumn.

Spinach makes a delicious baby leaf, either cooked or raw.

Harvesting and eating

Use a knife to harvest the leaves, taking them when they are young and small and before they have started to thicken up if you are cropping for salads or leaving them to grow slightly larger for cooking. Condition those to be eaten raw by leaving them in a bowl of water for a couple of hours. Leaves to be cooked will just need to be well washed. Spinach is cooked almost the moment it meets heat, so you can very quickly add interest and nutrients to a pan full of pasta and pesto by throwing in some baby spinach leaves and mixing them in. Once they have wilted, the dish is ready to serve.

Young spinach leaves are perfect in a potato and chickpea curry, or wilted and used as the base for a piece of fish or a soft poached egg. To wilt spinach, wash it and then put it into a pan with a knob of butter (spinach loves butter). Cover the pan and turn the heat to medium; the water on the leaves will be enough to steam it. Once it has collapsed, put it in a colander and squeeze it to release the excess moisture before chopping and serving.

Creamed spinach

This side dish is a fine use of baby spinach leaves – smooth, creamy and nutmeg-infused – but you can also use slightly larger leaves that are past their best for salads.

30g (2 tablespoons) butter
3 shallots, finely chopped
3 tablespoons plain (all-purpose) flour
180ml (¾ cup) full-fat milk
400g (14 ounces) spinach leaves
(enough to fill a large saucepan, pre-cooking)
60ml (¼ cup) single (light) cream
salt and pepper
freshly grated nutmeg

1 Melt half the butter in a saucepan and sauté the shallots on a low heat until translucent. Stir in the flour, cook for 2 minutes and then slowly start adding the milk, stirring, to make a roux. It should make a thick paste at first, which will thin out into a sauce as the volume of milk increases.

2 Meanwhile, wash and drain the spinach then wilt it with the butter in a saucepan over a medium heat. Once it has wilted, remove it from the heat and allow it to cool, then place it in a colander and squeeze out the excess moisture. Chop it and stir it into the sauce, along with the cream and salt and pepper to taste. Grate nutmeg over the top and serve.

New Zealand spinach

50 days

This is not a true spinach but has very similar, tender little triangular leaves that can be used in exactly the same way; the flavour is a little milder than that of true spinach. It has a sprawling nature and so is very useful as a ground cover for keeping down weeds over the summer. It grows perfectly happily through the hot conditions that often cause true spinach to bolt, so it's useful for filling that potential spinach gap and for planting in dry, sunny corners of the vegetable patch.

Cultivation

As these are sprawling plants you can sow them slightly differently from most cut-and-come-again crops, making a single sowing, planting them out as single plants and letting them trail across the ground to be picked all summer long. Germination can be slow, so soak seeds in water overnight before sowing. Sow indoors in mid-spring in seed trays and plant out when all danger of frost has passed, spacing the plants about 45cm (18 inches) apart. Alternatively, you can sow direct in late spring or early summer; sow a few seeds at each station and thin to one if all germinate. They should need very little watering. Although these are perennial plants in warm climates they are not frost-hardy and so will die off at the first frost. Pull them out and put them on the compost heap then sow again the following year.

Harvesting and eating

Pick the leaves off the stems when they are about 7.5cm (3 inches) long. This will encourage more leaves to be produced. Use them just as you would spinach: in cooking and raw in salads, where they have a mild, sweet flavour.

Grow large, individual plants of New Zealand spinach and then pick off the small triangular leaves from the stems.

Chard

70 days

Despite being an utterly dependable plant and a beauty in the garden and the salad bowl, chard is considered the poor relation of spinach. It doesn't have the same fine flavour, it's true, but it's fresh and green-tasting with a slight earthiness, and if grown as a perennial it's ready early in the year when other salad leaves are still only germinating. The glossy mid- to dark green leaves, sometimes purple-tinged, are nice enough, but the beauty is in the mid-ribs and leaf veins. In some varieties they are pure white, set off against the dark green, while in others they are deep red, orange and yellow.

Cultivation

Germination rates are great with these seeds and plants regrow so dependably after cutting that two sowings a year should be plenty to keep you going all year round. Make the first sowing in early spring in seed trays under cover or direct into the soil in mid-spring; a second sowing in late summer will give you winter leaves. Don't expect quick germination at either end of the year. In mild climates you can grow chard as a perennial for large leaves, but you won't get the same tiny tender leaves that you get with freshly sown plants, and the taste of the big leaves is not great when they're raw.

Recommended varieties

'Bright Lights' Commonly known as rainbow chard, this variety produces leaves in a range of greens and purple and leaf stems from red through orange and bright yellow to white. One of the most beautiful things you can grow in the vegetable garden.

'Canary Yellow' The tastiest of all the chards, this has clear yellow leaf stems and mid-green leaves. In the USA, this variety goes by the name 'Bright Yellow'.

'Fordhook Giant' This tasty variety has deep green crumpled leaves and broad, white mid-ribs.

'White Silver' Perfect white stems and mid-green leaves.

Chard 'Canary Yellow' and a red-stemmed leaf of multi-coloured chard 'Bright Lights'.

Harvesting and eating

The leaves grow as a small rosette and you can just pull off the outer leaves as they reach the size you want or use a knife to cut them all off at once. Use the tiny leaves raw to brighten up a salad and the larger leaves stir-fried, sautéed or steamed: ideally, strip the mid-rib from the leafy part and start cooking it a few minutes before adding the leaf. The leafy parts will wilt and the stems will retain their crunch for a lovely contrast in textures. Chard is great shredded in soups and stews or used as a side vegetable stir-fried with garlic and chilli.

Kale

42 days

Grown to maturity, kale makes a big, handsome, sturdy plant with tasty bluey-green, dark green or red leaves that look great in the vegetable plot all winter. It can also be grown as baby leaves, sown close together in guttering or straight into the soil and cut when they are tiny. Picked young like this they are delicious raw or just lightly cooked. They make a substantial and colourful addition to the salad bowl, with a crunchy texture and fresh, cabbagey taste.

Cultivation

Make a new sowing every few months, from late winter under cover. Through summer you can sow them direct into the ground outside, but make the late-summer sowings under cover again for cutting over winter. Three sowings over the season may be enough, as they take fairly well to cutting and quickly produce another flush of leaves. They will stand through winter under some cover and make the perfect base for hearty winter salads. The last sowing of the year should still be producing leaves during the following spring.

In summer, cabbage white caterpillars can be a problem on members of the cabbage family, but these leaves are picked while so young that the problem is unlikely to have time to develop. You will need to protect young plants from slugs.

Recommended varieties

Cavalo nero A type of kale from Italy, this has black-green pointed leaves with a great flavour. In the USA, it goes by the names lacinato, black Tuscan kale, nero toscana, black palm, and dino kale.

'Pentland Brig' Beautiful blue-grey curly-edged leaves with pale mid-ribs and veins. They are thin and so are particularly tender and suitable for cut-and-come-again treatment.

'Red Russian' Highly attractive blue-green serrated leaves with reddish-purple midribs and backs. They are mild-flavoured and smooth-textured, lovely raw when small.

Harvesting and eating

Start harvesting when the leaves are 5–7.5cm (2–3 inches) long, using a pair of scissors and cutting just the amount you want for maximum freshness. Condition in a bowl of cold water before spinning, drying and dressing. You can also stir-fry these leaves or steam them lightly and serve as a side dish with butter and salt and pepper. Kale is a wonderful source of dietary fibre, protein, iron, magnesium and vitamins A, B9 (folic acid), C and E.

Mark: "Interplant brassicas such as kale with nasturtiums - they cover the ground quickly, keeping moisture in and weeds out. They also act as a sacrificial plant for the cabbage white butterflies to lay their eggs on, leaving your brassicas relatively untouched."

Stir-fried kale with garlic and chilli

several large handfuls of baby kale
2 garlic cloves, minced
1 chilli pepper, deseeded and finely diced
sunflower oil, for frying

Leave the kale leaves whole if they're small, or chop them a little if they're larger. Heat the oil in a wok then throw in the garlic, kale and chilli pepper and cook for a few minutes, stirring, until the kale softens. Serve immediately as a side dish with noodles.

Lettuce

21 days

Lettuces are the mainstay of the salad bowl, and so they should be: they may not offer the gastronomical fireworks of some of the other plants in this chapter, but their leaves can contain sweet or bitter notes and can be crisp or soft to the bite. They are produced in a great range of colours, textures and shapes and if you look beyond the more obvious varieties you can use lettuces themselves to liven up salads. But their main role is as a foil for other flavours and colours, and this they perform admirably, acting as a neutral receptacle for dressings and other adornments.

Cultivation

The most common way of growing lettuces is to plant them out singly and allow them to grow a head, which means that the leaves fold in on themselves and on maturity you harvest the entire plant – a round, more or less hard ball of inner leaves blanched pale and sweet through lack of light, surrounded by fuller-flavoured and slightly tougher outer leaves. It's a fine way to grow lettuce but it has its drawbacks. For one thing, more mature lettuces are fairly large, and although they are at their best eaten fresh, you may find half the lettuce is stored away for the next meal. Also, if you plant out a row of these lettuces at the same time you will find that they mature all at the same time. It's a classic glut plant, and it's hard to fully enjoy something when you're getting too much of it.

Instead, try sowing a lot of seeds over a wide drill and cutting when they are small, taking only as much as you need. Lettuces can take two or three cuts before they start to fade away and need to be replaced. With a little planning it's possible to eat baby lettuce leaves all year round, since there are several varieties that have been specially bred to grow well in winter.

Lettuces don't like too much heat and you may struggle to get them to germinate at the height of summer. Try pre-watering the seed drill to cool it down or keeping seed in the fridge for a couple of weeks prior to sowing. Plants will also try to bolt in hot weather, putting all of their energy into flowers and seeds rather than into leaves, and the leaves that they do produce can be bitter and coarse. The shady, awkward spot in your vegetable patch is the place to grow your lettuces, and if you have no shade, try growing them on the shady side of rows of climbing peas or beans. Cutting regularly and watering well in dry spells will help to avert bolting disaster. Cool, dull summers will at least come with the consolation of a great many perfect lettuce leaves.

The trickiest thing about lettuce is pacing your sowing so that you're never overwhelmed or lacking. A steady supply is the goal, so that you'll always be eating them when they are little and sweet and never making do with slightly larger leaves. Sow every few weeks – if you find it hard to remember to do this, set yourself a rule to make a fresh sowing before you take the first cut of any one batch of leaves, then you'll always have some bringing up the rear.

Lia: "In hot weather I sneak out to the vegetable patch in the early evening to sow my lettuces. They have a need for a cool period about three hours after sowing, and so I get far better germination from these twilight sowings than I do from those made in the day."

Sow lettuces close together to harvest as cut-and-come-again leaves.

Recommended varieties

'Freckles' A romaine-type lettuce which has particularly crispy and sweet leaves, 'Freckles' lives up to its name and is dotted all over with reddish splotches. The leaves are glossy and crinkled and the effect is very attractive. Red lettuces need to be grown in good light if they are to produce their full colours; in low light they'll remain green, or their red markings will be very pale. 'Freckles' is slow to bolt, which makes it a good variety to grow for summer leaves. It's fairly hardy so good for early spring and late summer sowings as well, although it may not last through winter.

'Green Oak Leaf' This is perhaps the best lettuce for cut-and-come-again cultivation. Its habit – a loose centre with shapely individual leaves – is perfectly suited to it and even if left to mature it will only ever form a very loose heart of leaves, which can still be plucked from the plant individually. The leaves themselves are mid-green and wavy-edged, reminiscent of an oak leaf, unsurprisingly. They are tender of texture and mild of taste and make a good salad base. Sow from early spring through to late autumn.

'Reine des Glaces' This beautiful pale green lettuce is an iceberg type, so the glossy leaves are grown for their crisp and juicy texture rather than their flavour – it's among the best lettuces for cooking. The edges of the leaves flick out, pretty and flame-like. It's a lettuce that loves to heart up, so do grow a few as individual plants as well as for cut-and-come-again leaves. It's fairly resistant to bolting, so can be sown from early spring to early autumn.

'Winter Density' A great lettuce, mid-green, sweet and crunchy and tolerant of both heat and cold. It does well throughout the winter and is one of the best for late and early sowings. Make early autumn sowings for eating throughout winter and early spring. Sow under cover for winter if you can, otherwise it should do fine under a covering of horticultural fleece out of doors.

Harvesting and eating

When the leaves are 5–10cm (2–4 inches) high, use a pair of scissors to cut them about 1cm (½ inch) or so from the ground, taking care not to damage the growing tip. They are possibly the leaves that will benefit most from conditioning in a bowl of ice-cold water for a couple of hours, which allows them to plump up and become crisp and crunchy. Spin and then pat dry thoroughly before adding salad dressing and eating immediately (if you dress them before you are ready to eat them, the leaves will go limp again). If you don't eat all of the leaves at once, they will keep for a few days in a sealed plastic container or bag in the fridge.

Lettuce leaves are surprisingly good cooked, the leafy parts wilting to soft and yielding, the stems retaining a little crunch. Petits pois à la française (sautéed onions, peas and lettuce cooked in stock) is a great dish for dealing with a lettuce glut.

There is a great variety of prettily shaped and coloured lettuces; mix up your sowings so you always have a selection to choose from.

Sorrel

50 days

This little arrow-shaped leaf brings sharp, lemony bite to the table. A perennial, it's most often grown as an annual in order to keep the leaves small and tender, but either way it's extremely easy to care for. This is not a plant to use in great abundance: its sour taste comes from high levels of oxalic acid, which is not ideal eaten raw in large quantities. It's also rich in vitamins and minerals. As a sharpener among other salad leaves, or cooked down into a deliciously citrusy sauce, it makes a wonderful and unusual ingredient. You may find wild sorrel growing in roadside verges, but it's not as delicate-tasting as the improved, cultivated varieties, which are well worth seeking out and growing for yourself.

Cultivation

Sorrel doesn't need full sun and isn't too fussy about soil; sun or partial shade in most soils is fine. It's a hardy perennial and you can grow it as such, sowing seed in early spring, planting out small plants in late spring and then harvesting small leaves from them all year round. For really small and succulent leaves, though, it's best to sow it regularly and treat it as an annual. Make three sowings a year, one in early spring, one in early summer and one in late summer, broadcasting seeds into a wide drill so that they grow into a mat of foliage to cut from. You'll need to water while the seedlings are establishing themselves, but you should find sorrel fairly drought-resistant later on.

Grown as a perennial, sorrel is particularly useful in your garden as it's deep rooting and draws up nutrients from the lower levels of the soil. Cut the leaves regularly even if you aren't going to eat them and use them to mulch other plants. They will keep down weeds for a while and then slowly rot down, releasing those nutrients as they go. You can also put these unwanted leaves onto the compost heap to enrich your compost.

Recommended varieties

'Blood-veined' A mid-green leaf with deep red-purple leaf stalk and veins, this looks beautiful in the vegetable patch. It has the same lemony flavour as the other sorrels. In the United States, this goes by the name 'Red Veined Sorrel'.

'Profusion' This is a non-flowering variety, so you'll never get a problem with bolting or with the plant self-seeding.

'Silver Shield' Also known as buckler-leaved sorrel, this is a great improvement on wild sorrel, with a more delicate flavour and more succulent, tender leaves. It's named after the shield-like shape of the leaves. While this isn't the most productive sorrel it's not a crop that you will need to harvest a great deal from, and the improved texture and flavour more than make up for the lack of mass.

Harvesting and eating

Harvest leaves when they are about 12.5cm (5 inches) high, using scissors to cut them about 2.5cm (1 inch) from the ground so that the growing point is untouched. Use sparingly in salads, just to add a little zesty zing. If you have an abundance of leaves, try cooking them; they make an amazing sauce, marry happily with eggs and cheese, and also make a beautiful soup.

Buckler-leaved sorrel is one of the tastiest and most delicate types.

New potatoes with sorrel sauce

This sauce requires almost no cooking and coats the potatoes in a delicious sauce, savoury and buttery.

pinch of salt
10 baby new potatoes
175g (1½ sticks) butter
several handfuls of sorrel

Bring a pan of lightly salted water to the boil, add the potatoes and cook until tender (about 15 minutes). Drain the potatoes and return them to the pan, then immediately throw the butter and sorrel in with them and return to the heat. Keep stirring over a low heat for a few minutes until the sorrel turns khaki-coloured and starts to dissolve into the butter.

Erbette

30–40 days

This traditional and delicious Italian salad leaf falls somewhere between lettuce, chard and spinach in taste and texture and can be used as you would any of those: raw in salads or wilted in savoury dishes. It's very reliable and attractive and an unusual alternative to either baby spinach or lettuce as the mild base for a salad. It also seems very reluctant to run to seed.

Cultivation

Erbette grows well in sun or partial shade and is very happy to be cut back several times, regrowing strongly. Sow seed fairly densely in a wide drill drawn by the back of a rake and water well, making sowings about once a month from early spring under cover for leaves from early summer through to autumn. You can make final sowings under cover but erbette isn't especially hardy and is best grown for spring, summer and autumn.

Harvesting and eating

Cut when the leaves are 5–10cm (2–4 inches) tall, leaving the base to regrow. Condition in a bowl of ice-cold water for two hours before serving as part of a salad. Erbette can also be cooked like spinach (see page 119).

Chop suey greens/ Garland chrysanthemum

21 days

Also known by the cultivar name 'Shungiku', chop suey greens are a type of chrysanthemum and produce delicate, finely cut, silvery-green leaves with a subtle and aromatic flavour, leaning towards bitter. Left to mature they form small, bushy plants from which young leaves can be plucked, but the bitter edge becomes less pleasant as the plant ages, so they are best eaten young.

Cultivation

Sown in wide drills, they will make a swathe of feathery, silvery leaves. Make regular sowings from mid-spring onwards, direct into the ground, or start earlier under cover. To enjoy them at their youthful best you should make fresh sowings every few weeks. They will need to be kept well watered and prefer a spot in partial shade. If you let any plants get too large for leaves, allow them to produce their beautiful flowers, which are two-tone yellow and white and will make a splash in the vegetable garden. They are also edible and are used in a Japanese pickle.

Harvesting and eating

The smaller you pick the leaves the milder they will be; the bitter flavour makes a fabulous addition to dishes, but is best when it's not too strong. Start cutting when the leaves reach 6–10cm (2½–4 inches) in height. Add them in moderation to salads and sushi or steam or stir-fry them and add them to Japanese and Chinese dishes.

Claytonia/Miner's lettuce

42 days

A succulent, mild leaf, claytonia is grown specifically for winter leaves. It is brilliantly hardy and productive throughout winter, surviving well out of doors but producing even more perfect leaves under a little cover. Its common names are winter purslane and miner's lettuce, the latter referring to its use by California gold-rush miners who foraged it and ate it to prevent scurvy. While it's a North American native, it's more popular as a salad plant in Europe than in the United States. The leaves are mid-green and diamond-shaped if kept small, but if the plants are allowed to grow beyond that stage the leaf shape changes until it becomes almost perfectly circular, around the stem. In spring the small clusters of edible white flowers appear surrounded by this circular ruff-like leaf, and they can be eaten too. Claytonia has a slightly tart, spinach-like taste, though not as strong as spinach.

Cultivation

Claytonia naturally grows as a loose rosette of small leaves, so rather than sowing thickly as with other cut-and-come-again leaves, you will be aiming for individual plants. Sow in mid- to late summer for autumn and winter crops. Ideally you want to sow thinly, but the seed is particularly tiny so this can be tricky. Sow a row and then thin out plants to 7.5cm (3 inches) apart (eating the thinnings, of course). Start picking when the stems are 12.5cm (5 inches) tall.

You can also make spring sowings for summer leaves, but as there are many other leaves available for summer, claytonia is best sprung on the world when all the others have faded away.

Harvesting and eating

Claytonia isn't especially strong in taste but it's great for adding a juicy texture to salads of tougher winter leaves. Start harvesting when the leaves are still tiny. Unlike most of the salad leaves, the advent of flowering doesn't reduce the quality of the leaves – in fact they are at their most succulent and delicious when the little clusters of white flowers appear. Backed by the little circular leaves, they look great in an early spring salad. You can also cook claytonia and use it as you would spinach (see page 119).

Rocket/Arugula

28 days

Spicy, piquant and intensely peppery, rocket is a brilliant leaf for growing small and cutting often. Although its strength makes it a good addition to a mixed salad of milder leaves, it's interesting enough to make a salad of its own as well, and not too overwhelming with it. Something about the taste of rocket seems to enhance other flavours, bringing out their savoury notes, so grow and use it in abundance. A handful strewn across the top of a pizza after cooking is delicious.

Cultivation

Rocket takes beautifully to classic cut-and-come-again treatment – sow thickly and cut regularly once the leaves are 7.5–12.5cm (3–5 inches) high. They will come back well but will start to toughen after two or three cuttings, so start a fresh sowing when you take the first cut.

Make early sowings under cover in late winter and the first outdoor sowings in early spring. Rocket is a cool-season crop which struggles a little in summer, when it will quickly run to seed. You can try to minimize this with regular cutting and watering, but crops will always be at their best and easiest to manage in spring and autumn. It can grow very well through winter from a late-summer sowing, particularly under cover.

Flea beetle can be a big problem on rocket, marking the leaves with little dots. The damage is only superficial, but can be offputting. Grow under horticultural fleece if it bothers you.

Recommended varieties

Salad rocket, *Eruca vesicaria* subsp. *sativa* (often sold in the UK as *Rucola coltivata*), is the milder-tasting form, with slightly larger and smoother leaves than wild rocket (*Diplotaxis tenuifolia*), which has a more pungent taste and more deeply cut leaves. Wild rocket will last longer without turning tough, but this plant isn't as productive as salad rocket.

Deliciously savoury rocket leaves are at their most enticing if they can be kept clean of flea-beetle holes.

Harvesting and eating

Cut the leaves when they're small and tender, using a pair of scissors and leaving the growing point intact. Use in mixed salads or as a salad in itself, or as a leafy vegetable in risottos.

Rocket pesto

2 tablespoons walnuts
several large handfuls of rocket leaves
pinch of salt
25g (¼ cup) grated Parmesan cheese
25g (¼ cup) grated hard goat's cheese
120ml (½ cup) extra virgin olive oil, plus extra for finishing
1 garlic clove, crushed

1 Preheat the oven to 190°C/375°F/Gas 5 and toast the walnuts for about 7 minutes. Remove from the oven and allow to cool.

2 Basil pestos can be made with a pestle and mortar but the stem of rocket is a little too tough for this. Snip off as much off the stem as possible then put all of the ingredients in a blender and whiz together. Pour into a jar and top with a slick of olive oil to help seal and keep fresh. Stir through freshly cooked pasta.

Quick-harvest vegetables

Grow crops fast and they'll be sweet, tender and on your plate early in the year. Cultivating your own vegetables is the only way to have them at this moment of perfection, and it's a huge privilege. If you were a farmer you'd have to concern yourself with keeping them in the ground for as long as possible in order to harvest big, weighty vegetables and maximize your margins, but you're not; it's your choice whether to have pleasant, medium- to large-sized carrots, just as you'd find in the shops, or go for tiny, sweet ones, just 10cm (4 inches) long. Often, upping the quality of a crop tenfold is just a matter of harvesting early, catching that sweetness and avoiding the watery and woody elements that creep in over time – so this chapter is dedicated to the commonplace vegetables that become a little bit extraordinary when they're picked very young.

Baby carrots, picked while young, sweet and tender.

Mark: "When I'm planting small seeds I mix them with sand – it helps to ensure even sowing and marks the area where I've sown."

SOWING TENDER ANNUALS

Many of the most delicious vegetables – tomatoes, courgettes, dwarf French beans and pickling cucumbers, for example – are tender: they won't tolerate frost at either the beginning or the end of the season. Despite this they all need a good long season in order to grow to maturity and start producing fruits, so gardeners extend the season by starting them off indoors. Some of them can be sown outdoors after the last frost and still bear fruit, but you will get earliest results if you sow indoors and harden off gradually as the weather warms.

Sow straight into a small pot, as most of these make large plants quite quickly. You can plant them out when all danger of frost has passed, but they will need a little time to 'harden off' and become accustomed to outdoor life. Start off with a few hours in a sheltered spot in the day, then bring indoors every night. Over several days gradually increase the time outdoors until finally they are outside around the clock.

SOWING HARDY PLANTS

Hardy seeds such as carrots, turnips, beetroots and radishes do not need such elaborate cosseting. They can tolerate a bit of cold, and even a little frost, and so can be sown where

they are to grow, which makes life easier all round.

Comparatively tolerant they may be, but they do have some requirements. You will need to take a little care over the soil they are going into. Above all it needs to be crumbly and well worked over, so that it is loose and friable, and water, air and roots can pass through it easily. After digging your seedbed over and breaking up any lumps, use a rake to roughly level the ground and the back of the rake to create a smooth surface. Cover the soil with a cloche or a piece of black or clear plastic for a few weeks so that the warmth of the sun will be captured. Then, when you do sow, it will be into dryer, warmer soil and germination will be quicker and more consistent as a result. Covering will encourage a flush of weeds to sprout, which you can hoe off with minimal disturbance just before sowing is done.

Sowing in an arrow-straight line takes some of the guesswork out of weeding in the early stages, before you can really tell your vegetable seedlings from weed seedlings. A garden line will keep you on the straight and narrow. There are many commercially available, but two sticks and a length of string do an equally fine job. Tie a stick to each end of the string and push them into the ground at each end of your row, so that the string is close to the ground and so gives a good guide. Draw the corner of a hoe or rake along the line to make a little ditch – generally speaking, seeds like to be sown to a depth equivalent to their size (for example, broad beans like to be covered by 2cm/1 inch or so of soil). Watering after sowing often scatters the seed, so water the ditch lightly first using a watering can with a fine rose.

Sow the seeds thinly along the ditch before covering them lightly with soil. As it's more or less impossible to sow as evenly and thinly as you need to, little clumps of seeds will germinate. Thin them out as they germinate and grow, so that plants are evenly spaced.

SUCCESSIONAL SOWING

If you sow one large batch of carrots, for example, at the start of the season, in 6–8 weeks you'll have an abundance of perfect and sweet baby vegetables – perhaps too many to eat immediately. And what happens to baby carrots if they're not picked and eaten immediately? All too quickly, they grow beyond their miniature prime, turning into medium carrots, then big carrots, losing that sweet and tender loveliness. For most households a steady, smaller supply suits best and for this you'll need to get into the habit of sowing small amounts at regular intervals, rather than in one large batch. This way you will never have more than you can eat maturing at any one time, so you can always harvest crops when they are at their best. Each time you sow one of the vegetables suitable for successional sowing, make a note in your diary or calendar a few weeks later to remind you to sow again.

Hardy annuals such as carrots and spring onions can be sown direct into pots. Sow seed across the top of the pot, cover with compost, and water in gently.

HARVESTING AND EATING

The trick to harvesting most of these crops while they are young and lovely is vigilance. Fast-growing crops can turn from tender to tough rapidly, so you need to keep on top of them if you're going to enjoy them young. Check them every couple of days, nipping them off the plant when they are at their finest. Likewise there's little point in picking them sweet and young and then letting them sit around and toughen up for a few days – pick just what you need and harvest often. Plants such as cucumbers and beans stop producing more fruits if those on the plant become large. The plant's aim in life is to reproduce – this is why it's making flowers and fruits – and if it thinks its work is done it slows down flower production, putting all its energy into ripening up the fruit rather than generating more. It's another reason to keep picking young.

To sow crops directly into the ground first work the soil to a fine, crumbly tilth and then use a line and pegs to mark out where they are to be sown. Sow as thinly as you can, cover with a little soil, water in and label.

Mark: “When I’m about to plant out a tray of seedlings, I sow another tray first. This ensures that I always have the next batch on its way.”

Cherry tomatoes

65 days

Tomatoes are a tease. You sow in spring, picturing yourself sitting out of an evening on a sun-warmed terrace eating monster slices of beefsteak tomatoes, dotted with lumps of mozzarella and drenched in olive oil. The terrace warms, then cools. The nights start to draw in, you abandon the terrace and then, finally, your tomatoes ripen. The large-fruited cultivars, grown outdoors, are very definitely a late summer and even early autumn treat. Even in warm, dry climates the tomato season can be short. High temperatures in midsummer make pollination erratic and can completely bring tomato production to a halt.

But there are tricks you can use to obtain earlier crops, the simplest of which is to grow cherry tomatoes. Tiny, sweet little things ripen much earlier than their big, beefy counterparts. Sweet, juicy and (if you choose the best varieties) full of flavour, cherries are equally perfect enjoyed raw and fresh from the plant or cooked. Another trick is to sow them as early as you dare and give them the longest possible time in which to fruit.

As well as getting you eating tomatoes on the terrace on a summer's evening while others lament their green beefsteaks, there's another reason to look for speed in tomatoes. Blight is a big problem in varieties grown outdoors. Spores are carried in the air and often hit just as tomatoes start to ripen, turning foliage yellow and then black, ruining fruits almost overnight. It's heartbreaking when you've been nurturing them all summer. Cherry tomatoes often beat it, not through resistance, but by ripening before blight arrives. They are the only type of tomato I grow out of doors for this reason. But to be sure to avoid the blight heartache, grow even cherry tomatoes under some sort of cover. That blackening of nearly ripe tomatoes is a tough one to bear with grace.

Cherry tomatoes such as 'Gardener's Delight' ripen earlier in the season than larger-fruited types.

Mark: "Late in the season, when the tomatoes are grown but not yet ripe, I prune off any leaves that are shading the fruit – the extra light and heat bring the fruit on much faster."

Cultivation

Getting your timing right is important when you are sowing tomatoes. They need a long growing season, yet you make problems for yourself if you sow too early. Lucky gardeners with greenhouses can sow as early as the end of February, but those growing first on windowsills and then outdoors need to wait a little longer. Find out the last expected frost date for your area and count back six weeks from that. This is about the time it will take for your plants to germinate and grow big enough to plant out; sow earlier and they will get stretched and leggy in low indoor light, and then struggle with outdoor conditions. Once frosts have passed and the plants are large enough, harden them off by putting them out for a few days during the daytime only, and then plant out.

Tomato growing is a race to ripening. This isn't a plant that will endlessly go on and on producing: once it gets cold, it's over. You have a finite amount of time, so you need to concentrate the plant's efforts on producing a finite number of fruits. There are two types of tomato: determinate and indeterminate. The former are also known as bush types, which gives you an indication of their habit. They have multiple branches and cover themselves with fruit, but they know when to stop branching and when to start fruiting, so they don't need any kind of pruning. They do need starting off early, though, or you will be left with a bush full of unripe tomatoes at the end of the season.

Indeterminate types are a slightly different story. They are also known as vine or cordon types and much like a vine they have no idea when to stop. They will keep on growing and trying to produce more and more tomatoes. At the same time they will try to send out side shoots. The plant is trying to do more than it easily can in a season, and both excesses need to be curtailed. Nip out the side shoots as soon as you see them and remove the top from the plant once you can see that four or five trusses of fruits have set. This will stop the plant from reaching ever upwards and setting ever more fruit and force it to concentrate on ripening those that it has already set. All the plant's energy will then be poured into making those few fruits delicious.

Tie the main stem of your tomato plant to a sturdy stake

Recommended cherry tomato varieties

'Black Cherry' (cordon) Dark and exotic-looking yet sweet and juicy. It looks great mixed with yellow and orange varieties.

'Gardener's Delight' (cordon) This straightforward cherry tomato does everything necessary: small, sweet and abundant fruits are borne on long trusses.

'Orange Santa' (cordon) A cherry plum with long, orange, very sweet fruits.

'Roma' (bush) A red cherry plum or grape tomato, with a sweet and slightly spicy flavour.

'Sub-Arctic Plenty' (bush) Possibly the earliest-ripening tomato there is. First developed during World War II to provide the US Air Force in Greenland with fresh tomatoes, this variety can produce tomatoes even in cold conditions, and therefore fruits very quickly in warm months.

'Sungold' (cordon) If you are after sweetness then this is the tomato for you – it's one of the tastiest by far. The fruits are a beautiful yellow-orange colour when fully ripe.

Harvesting and eating

Cherry tomatoes are a little different to most of the crops in this book in that there's nothing to be gained by harvesting them early – they'll taste thin and sharp compared to the rich complexity of those that are fully ripe, so leave red varieties until they turn bright red, then dull a little. Yellow varieties are harder to judge and you may taste a few sharp ones before you get a sense of when they are ripe. You want no hint of green, but a little of orange.

Eat cherry tomatoes fresh from the plant, warmed by the sun, and popped straight into your mouth as the ultimate garden snack. Mix reds, yellows and different shapes and sizes to make a particularly pretty salad. Roasting sweetens them further and they are a tasty addition to roasted Mediterranean vegetables, thrown in at the later stages.

Lia: "At the end of the season I put green tomatoes into a paper bag and store them in a dark place for a few weeks. They ripen beautifully there. Ripening on a windowsill toughens the skins."

Oven-dried tomatoes

The trouble with tomatoes is that they're prone to gluts. By oven-drying them and preserving them in olive oil you can enjoy them several months after the end of the season. You are essentially just dehydrating them slightly for storage, but this also intensifies the flavours. This recipe works for tomatoes of any size, but larger ones will need to be quartered and don't look as pretty as halved cherry tomatoes. It looks lovelier still made from a mixture of red, orange and yellow cherry tomatoes. Use small jars to store them in as they will go off quite quickly once opened – small Kilner or canning jars are best.

tomatoes
olive oil

1 Preheat the oven to 150°C/300°F/Gas 2. Slice the tomatoes in half and place them facing upwards on a baking tray. Put them into the oven and wedge the oven door slightly open to allow moisture to escape. Leave for 4–5 hours, checking occasionally. The tomatoes should dry out but not turn brown at the edges.

2 When the tomatoes are wrinkled but still juicy in the middle, wash the jars and dry them in the oven for at least 20 minutes.

3 Warm the olive oil in a pan. Place the tomatoes into the warm jars and pour the olive oil onto them, shaking a little to remove air pockets. Seal immediately. Eat in warm autumn salads or use to enrich a pasta sauce.

Early potatoes

70–90 days

There are four types of potatoes: first earlies, second earlies, maincrops and late maincrops. First earlies and second earlies produce fairly modest yields of soft-skinned potatoes, and they do it quickly: first earlies in about 10 weeks and second earlies in about 13 weeks. Maincrops and late maincrops produce larger yields of harder-skinned potatoes that store well, but take longer to mature (around 15 and 20 weeks respectively).

There are many reasons to restrict yourself to growing only early potatoes, and only in containers. One is that they are quick in and out of the ground, a gift for the impatient and those with crops they want to plant later. But this isn't just about speed, for when the time from planting to harvesting is short, the potatoes produced are little and tender. They may be a great source of carbohydrates, but these are not your boring bulky everyday potatoes; they are the kind that you would pay a premium for in the shops, small, waxy, sweet, and a world away from those that stay in the ground all season. Maincrop potatoes are big and floury, and above all cheap to buy – so buy, don't grow. They suffer from blight and scab later in the year, and slugs make holes in them as the growing season wears on. Quick in-and-out container-grown new potatoes don't suffer any of this: the quicker you grow them, the less time the many pests and diseases that stalk potatoes have to get a grip, and the better quality crop you get. Everyone wins, except for the slugs.

Cultivation

One of the loveliest things about potatoes is that you start them before the growing season is properly under way when nothing else is showing even a hint of ever wanting to grow again. If you get the urge to plant something at this time, make it potatoes. Everything else will do better if you wait.

Potatoes are bought as 'seed potatoes' – a confusing name as they are actually small tubers, mini potatoes that will sprout and grow other potatoes along their roots. In late winter, place them in trays somewhere light and cool but frost-free and allow them to 'chit' or sprout. Once the sprouts are about 5cm (2 inches) tall they can be planted. If you grow them in a container, growth will be quick, as potting soil is far warmer than the soil of the vegetable plot in early spring.

Potatoes are sensitive to frost, so don't plant them out too early. However, you will be covering the tubers with potting soil and this will protect them to an extent, so a light frost isn't a problem. Take a large pot or bag (even

a thick bin liner) and fill it with about 10cm (4 inches) of potting soil. Place a maximum of three tubers to a container and cover with another 10cm (4 inches) of soil. Water well. As the shoots show through, cover them with more soil, on and on until the soil reaches the top of the container. Water regularly and generously, feed at least once a week (liquid seaweed has the perfect balance of nutrients) and you will get armfuls of the loveliest tubers you have ever eaten.

Place up to three tubers into a small amount of soil in the base of a large pot, cover with a little compost, then water and wait for the shoots to appear.

Recommended varieties

'Anya' (second early) A pink-skinned cross between 'Desiree' and 'Pink Fir Apple', it has much of the latter's delicious nutty taste and waxy texture but none of its knobbliness.

'Arran Pilot' (first early) A waxy-fleshed potato with excellent flavour. It prefers a good soil to do well.

'Belle de Fontenay' (second early) A French potato, kidney-shaped, quite small, waxy and hard to beat for salads. It even improves with storage.

'Edzell Blue' (second early) If you're after something different and delicious, try this variety with blue skin. A very floury potato, almost round in shape, it's great mashed, baked, roasted or steamed.

'Purple Viking' (second early) A beautiful potato with deep purple skin splashed with pink. The snow-white flesh has a smooth texture and great flavour.

'Swift' (first early) This potato lives up to its name, as none are ready earlier. It's a floury variety that tastes and looks fabulous.

'Yukon Gold' (second early) With yellow- and pink-skinned oval tubers and buttery yellow flesh, this variety is very much at its best when freshly harvested.

Harvesting and eating

To harvest, just dig about below the plant with your hands 8–12 weeks after planting. The earliest ones will be luxuriously little and perfectly smooth, but you will get a more satisfying yield if you leave them a while longer. If you are growing them in a bag, you may be able to cut into the bottom and harvest the largest tubers, leaving the tops in place to keep on growing.

Put your potatoes into a pot of boiling water until tender – this may be as little as eight minutes – and serve with butter. New potatoes are delicious with a sorrel sauce (see page 136).

Lia: "I grow a few pots in my greenhouse. The extra warmth makes them grow particularly quickly, and produces even earlier, sweeter, and more succulent potatoes."

Harvest pot-grown potatoes while they are tiny and tasty – this method is not for bulk.

Baked, sea-salted new potatoes

Though perfect just boiled, new potatoes take on a different character when baked whole with crispy, salty skins.

new potatoes
olive oil
sea salt
sprigs of thyme

Preheat the oven to 200°C/400°F/Gas 6. Put the potatoes in a high-sided baking tray and drizzle with olive oil, turning them with your hands until all are coated. Sprinkle on plenty of sea salt and throw on the thyme, then bake for about 45 minutes. The insides should be soft and the skins crispy.

Radish

22 days

Radishes are perhaps the ultimate quick-growing veg. Sow them from mid-spring and they can be ready within a month – and if you pick them early, before they get too large, you'll avoid any trace of woodiness and the peppery heat will be lively yet mild, rather than eye-watering. Throw them whole or thinly sliced into a salad to add vitality and to benefit from their high levels of vitamin C and potassium.

Cultivation

This really is as simple as vegetable-growing gets. Thinly sow a row any time from mid-spring to early autumn and up come shoots within days, the little bright red edible crop maturing within just a few weeks. They prefer sun and a moist but well-drained soil, but aren't too fussy.

Because they take up so little space and are so quickly in and out, radishes make great catch crops for sowing in between those crops that are in the ground longer. Their fast germination also makes them useful markers for slow-to-germinate crops such as parsnips, a particularly handy technique if you are short on space. Again, they'll be out of the way before the parsnips need the room to spread out, but they might just remind you there's something there in the meantime.

Radishes are brassicas and flea beetles love brassica leaves. Covering the row with horticultural fleece can keep them at bay. You'll also need to protect against slugs, which can wipe out an entire row overnight.

Recommended varieties

'Cherry Belle' A classic and particularly fast-growing radish, producing rosy-red globes with snow-white, crunchy flesh.

'French Breakfast' A fine old variety well suited to being picked small and tender: it can go woody if left to grow large. Oblong in shape, and particularly tasty.

'Rudi' Dark red skin and crisp flesh.

Harvesting and eating

Pull radishes out of the ground, run them under a tap and they're ready to eat. Try them whole, smeared with a little butter and sprinkled with sea salt, or sliced into a salad.

If you've managed to grow the tops without too much flea beetle damage, you can also use them as a salad leaf, tossed in vinaigrette.

Radish 'Cherry Belle' is a classic red-skinned, white-fleshed radish.

Mark: "Don't forget to use the leaves in salads or stir-fries – they're delicious."

Carrots

50 days

Carrots can be two quite different beasts, depending entirely on how you grow them. Sown from mid- to late summer, and grown on into autumn and winter, they are hearty and solid, requiring slow cooking and oil to release their mellow sweetness; grown quickly in spring and summer they are a little mouthful of crunch and sugar, as good pulled from the ground, brushed off and eaten instantly as they are in any recipe.

Cultivation

When the growing season is in full swing you can get carrots from sowing to salad size in six weeks. Sow a small row of carrots or a potful every three weeks through the summer to ensure a constant supply of little ones all season long. They do best in humus-rich, well-worked but firm soil, not too compacted and not too loose. While they are simple to grow – sow them in the ground and dig them up – they are greatly troubled by the carrot fly, a nasty little pest that lays its eggs on the tops of carrots. As the carrot grows, the larvae hatch and burrow into the root, leaving tunnels through the tops. Carrot fly is attracted to the carrots by the fragrance of the leaves, so you can try companion planting with something even more aromatic: any member of the onion or garlic family makes a great partner plant. Thinning the carrots is a danger point, since the crushed foliage emits a stronger fragrance, so if possible do this in the evening on a still day and remove the thinnings immediately.

Otherwise, the simplest way to deter them is to erect a physical barrier. Cloches offer early crops warmth and protection, and later crops can be grown under horticultural fleece to keep the fly at bay. A barrier of mesh fencing provides some protection, too, since the fly will raise its flight path and pass overhead.

Lia: "Carrots do well in large pots. They like the loose soil and the pot also lifts them just above the flightpath of their main pest, the low-flying carrot fly."

Baby carrots can be eaten straight from the soil.

Recommended varieties

'Chantenay Red Cored' A tasty, deep orange, early maincrop carrot which grows quickly when sown in a warm soil. The roots are shortish and wedge-shaped so you won't need the deepest soil to grow them. Very reliable and high yielding.

'Paris Market' A globe type, short and round, and therefore good for heavy and compacted soils.

'Primo' One of the best for earliest sowings.

'Sytan' A variety that has shown good resistance to carrot fly.

Harvesting and eating

Pull baby carrots from the ground, wash off the dirt and eat raw. Alternatively, steam for a few minutes and toss in butter. You can also try them roasted whole with honey and herbs in a tray of other baby root vegetables.

Push a fork beneath your row of carrots to loosen the soil, and you can then gently pull small bunches of the carrots from the ground.

Carrot crudités with creamy dip

bunch of freshly harvested baby carrots
2 heaped tablespoons cream cheese
1 heaped tablespoon sour cream
small bunch of chives, snipped
1 garlic clove, crushed

Chop the leaves off the carrots, leaving 1cm (½ inch) of green, and scrub them clean. Mix all the other ingredients together to make the dip, reserving a small amount of chives to sprinkle on the top.

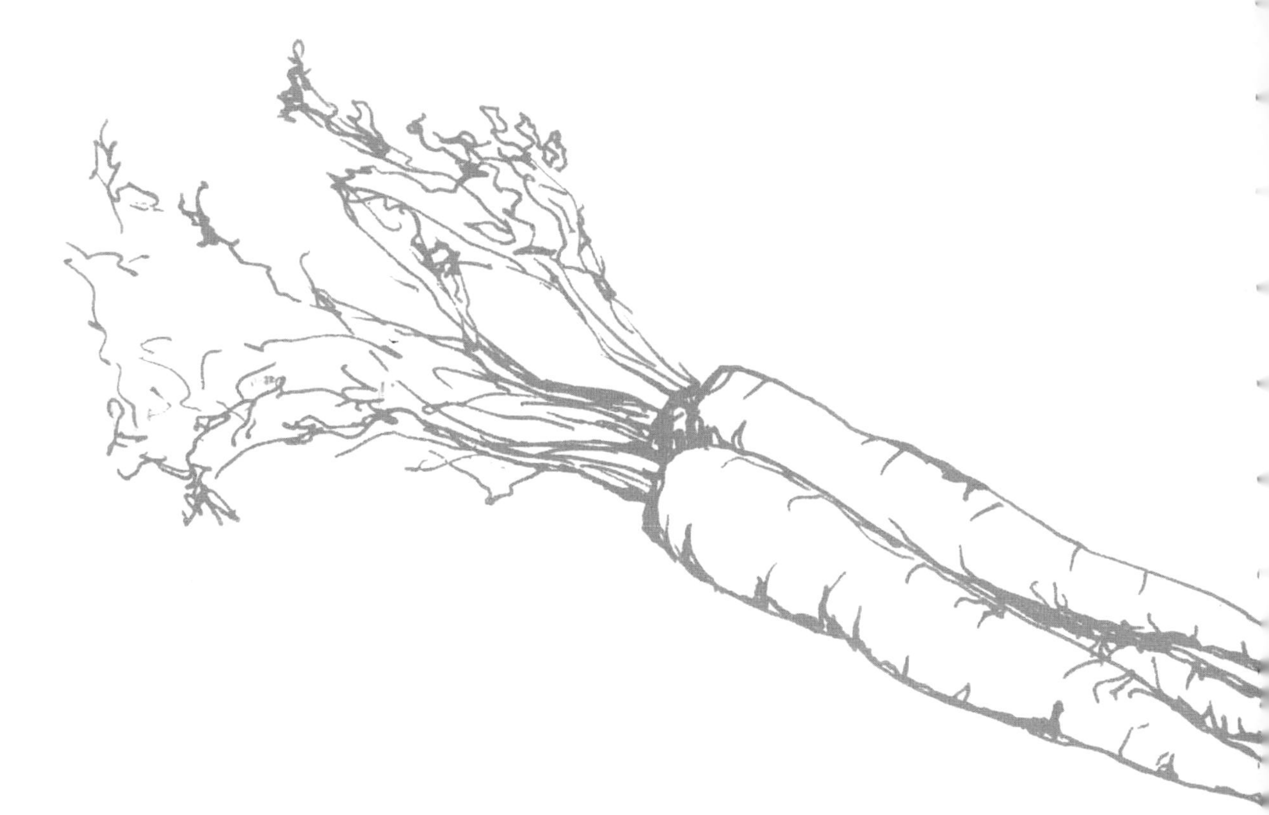

Eat small carrots raw or lightly steamed or roasted.

Courgettes/Zucchini

35–40 days

The essential summer vegetable, tender and tasty and one of the easiest to grow, courgettes will keep pumping out fruits all summer long once they get under way. There's a world of difference between a tender little courgette and a big watery one, though. Regular picking and temperance at sowing time are the keys to enjoying tiny courgettes at their sweet, glossy finest all summer long.

Cultivation

Courgettes are tender, so follow the advice on sowing tender annuals on page 150. They are such massively productive plants that it's important not to get too carried away when sowing. If you have too many plants you'll have too many courgettes, which means you'll get tired of them and stop harvesting them so often, and then they'll grow big and past their best. The trick with courgettes is always to leave yourself wanting, as that way you'll pick them when they're tiny. With this in mind, sow only four seeds at a time: assume one won't germinate, one will be eaten by the slugs, and two will be left for you. A few weeks after sowing your first few, sow a second batch, but again, a maximum of four plants.

Plant out into a planting hole that has been enriched with well-rotted manure or garden compost: courgettes are thirsty and greedy plants. Protect from slugs in the early stages, and water during dry times.

Lia: "Courgettes are very thirsty plants and produce best if kept really well watered. I make a shallow depression in the ground and plant into that. It means every drop of water that falls near the plants runs straight to the roots."

Small courgettes are firm and sweet, without the wateriness that comes as they mature.

Recommended varieties

'Alberello di Sarzana' Light green fruit with a slightly bulbous end and a delicate flavour.

'De Nice à Fruit Rond' Also known as 'Rond de Nice', this produces little round pale green courgettes, pretty and tasty.

'Nero di Milano' A wonderful Italian variety with very dark green skin. Very productive with an open plant habit that makes it easy to harvest and keeps the likelihood of disease low.

'Raven' Dark green fruits with particularly tender flesh are borne high up on neat, bush-style plants that don't sprawl, making harvest easy. Good for containers.

'Soleil' A bright yellow, glossy and sweet variety.

'Striato di Napoli' Productive yet compact plants that produce lovely striped green fruit.

'Zephyr' Bi-coloured and beautiful; the pale yellow fruits look as if they have been dipped in green wax at the blossom end. They also have an unusual nutty taste and firm texture.

Harvesting and eating

Harvest regularly and while the fruit are just 10–12.5cm (4–5 inches) long to enjoy them at their best and to encourage new fruits to follow on. As soon as you let a fruit grow large, production of new fruits slows. Use a sharp knife to cut close to the plant. Examine plants carefully for bounty, as young fruits can look very like stems, and it's easy to miss one. Eat immediately for the sweetest, mildest taste, though they will store in the fridge for a few days. Slicing them diagonally or into tiny rounds and putting them into a salad, raw, or cooking them quickly on barbecue skewers are just two of the many ways to enjoy them.

Harvest when the courgettes are about 12.5cm (5 inches) long, using a sharp knife to cut through the end closest to the plant.

Chargrilled baby courgettes with mint yogurt dressing

Courgettes love almost every herb, but mint is an especially summery accompaniment.

handful of baby courgettes
1 tablespoon olive oil

For the dressing

3 tablespoons Greek yogurt
1 tablespoon lemon juice
1 garlic clove, crushed
handful of mint, chopped
salt and pepper

1 Mix all of the ingredients for the dressing, leaving aside a small amount of mint.

2 Cut the courgettes lengthways into 1–2cm (½–1 inch) slices, coat with olive oil and place on a hot griddle. Cook for a few minutes, then turn and cook on the other side. Pile onto a plate and cover with the dressing, a little olive oil, and the remainder of the mint.

Dwarf French beans

60 days

Runner beans may be vastly productive and borlotti beans are certainly wonderful slow-cooked in stews, but if it's the youthful, the sweet and the tender that you value, French beans are the beans for you. Borne on small bushy plant, they are quick to mature and are particularly sweet and tender when picked young. They are one of the few beans (other than baked) that kids will eat with relish.

Cultivation

Dwarf French beans and climbing French beans produce much the same bean, but the dwarf version does it without you engaging with the fuss of building supports and tying up stems. It makes neat little mounds of foliage and pumps out beans at an astonishing rate: possibly not quite as productive as the climbing type but you'll certainly never be left wanting.

Though seedlings are sensitive to frost you can plant direct in mid-spring as long as you cover them with cloches. Sow seeds 30–40cm (12–16 inches) apart, placing two at each station in case of failures – you can pull out the smallest seedling if both germinate. Remove cloches after all danger of frost has passed. Make small sowings of four or five plants at a time every few weeks until late summer.

Recommended varieties

'Delinel' Large crops of smooth, green beans.

'Purple Teepee' Colourful purple pods borne above the foliage, which makes spotting and picking them easier.

'Rocquencourt' Pale yellow curled beans with a hint of green. This variety copes well with cold so is a good one to start early, under a cloche.

Harvesting and eating

Once plants start producing you'll need to pick them over every few days. This way the beans are always small and tender, and the picking encourages the plant to keep on producing for the longest possible time.

Eat raw or cook lightly, so that they retain their crunch. Steam for a few minutes and toss in butter, or cook them in butter with some crushed garlic.

Grow yellow, green and purple dwarf French beans for colourful dishes.

Pick French beans when they are small; they get tougher as they mature.

Salade Niçoise

To keep the beans in this (and any other) salad crunchy, blanch them for 1–2 minutes, drain and place immediately in cold water for 1 minute before draining again.

Serves 4 as a starter

1 medium-sized tuna steak
2 handfuls of yellow and green French beans
4 tomatoes
8 new potatoes, cooked and quartered
4 eggs, boiled until just set (around 4 minutes) and halved
2 handfuls of lettuce leaves
1 red onion
6 anchovy fillets
black olives
a few sprigs of basil, shredded

For the marinade

3 tablespoons olive oil
1 tablespoon cider vinegar
6 chives, finely snipped
2 stems of parsley, leaves only, finely chopped
2 garlic cloves, minced

1 Combine the ingredients for the marinade in a jar and shake well.

2 Place the tuna in a flat dish and pour the marinade over, reserving some to use as dressing. Leave in the fridge for about 2 hours then cook on a griddle for a few minutes on each side. When cooled slightly, slice, combine with all of the other ingredients and pour the dressing over.

Cucumbers

40 days

Cucumbers have a bit of a reputation for being tricky greenhouse plants, but if you choose the right varieties you can grow them out of doors and with very little fuss. For quick crops there are several that produce small, almost round cucumbers, as sweet as they are crunchy. Gherkins are all that but, being smaller fruits, reach their goal even quicker than cucumbers.

Cultivation

Cucumbers are frost-tender, so follow the advice on sowing tender annuals on page 150. While this is essential for cucumbers, gherkins are so quick into production that you can sow after frosts are over and still get a great crop. Plant seedlings into the ground, protecting from slugs. They are climbing plants so will need a support. A trio of canes tied into a tepee works well and is the perfect support in a large container. Loosely tie in the stems on a regular basis and keep plants watered well.

Recommended varieties

'Burpee Pickler' A heavy yield of spined pickling cucumbers over a long period.

'Bush Champion' Short green cucumbers produced on a bush plant, which takes up much less room than vine types.

'Crystal Lemon' Small round tennis ball-sized yellow fruits are borne abundantly. They have deliciously sweet crunchy flesh.

'La Diva' (also sold as 'Diva') Long, slender, dark and spineless green fruits that are sweet and seedless. Grows well out of doors.

'Marketmore 76' Long, dark green fruit on a strong, healthy plant.

'Vert Petit de Paris' Tiny pickling cucumbers are produced prolifically on this reliable plant.

Harvesting and eating

Use a sharp knife to cut cucumbers from the plant after they have reached about 12.5cm (5 inches) long: the smaller the sweeter, and the thinner the skin. You can cut gherkins when they are tiny, just over 2.5cm (1 inch) long, and pickle as cornichons in tarragon and vinegar, or leave them to grow to around 7.5cm (3 inches) for larger sweet spiced pickles. Use them in home-made tartare sauce, or to provide the crunch in salads and burgers.

Lia: "A few years ago I invested in a heated propagator and I highly recommend it for germinating tender crops such as cucumbers and peppers. On the windowsill they can sit and sulk for a long time before germinating, but they can't resist the steady heat of the propagator and are usually up in days."

A young cucumber is just beginning to swell.

Spring onions

60 days

Usually milder than your average onion but still with a hearty bite, spring onions are the quickest onion you can grow. With a little planning you can have them fresh from the soil all year round, and there's such a range of dishes from salads to mashed potato that are enlivened by that strong but not overwhelming flavour you'll be glad to have a good supply.

Cultivation

Hardy and easy and with no particular garden foes, spring onions are a straightforward proposition. From early spring you can sow them direct into well-drained soil in a sunny spot. Sow sparsely, and once they have germinated thin to a final spacing of about 10cm (4 inches). Also try scattering seed thinly across a straight-sided pot. They will do well and look good, particularly the red varieties.

A small amount sown every few weeks through the summer will keep you well supplied, and they are small and neat enough to squeeze in between other crops; pop a row between rows of carrots as their onion scent can help to confuse the carrot fly. Make your final sowing in autumn, cover with a cloche and they'll be ready to eat in early spring. Sowings take around nine weeks to reach maturity, but there's no need to wait for them to reach full size. Start eating as soon as they look worth the trouble of pulling them up, whenever suits you.

Recommended varieties

'Deep Purple' Bright purple colouring and a more bulbous base than the other varieties, and grows larger.

'North Holland Blood Red' A red-bulbed spring onion. Those left in the ground will bulb up into full-sized onions, so harvest with this in mind, leaving space around those left behind.

'White Lisbon' The most widely grown and the best, crisp and strong-tasting.

Harvesting and eating

Once onions start to jostle each other, thin out and eat the thinnings, then just harvest as needed as soon as they're large enough. Eat raw, sliced and sprinkled onto the top of salads or noodles, or briskly stir-fry with other quick-cooking vegetables. Slow-braising them whole brings out a mellow sweetness that turns them into great side dishes.

Spring garden tart

Spring onions are so often a supporting ingredient, but, cooked gently to bring out their sweet side, they can be the centrepiece in a tart such as this. All sorts of early-summer green vegetables rub along beautifully with each other when combined in a tart. Use whatever vegetables and herbs are at the peak of freshness in your garden.

Serves 6

75g (⅓ cup) chilled butter
125g (1 cup) plain (all-purpose) flour
pinch of salt
1 egg, separated
a little cold milk

For the filling
20g (1½ tablespoons) butter
20 spring onions
large handful of spinach leaves, washed and roughly chopped
130g (1 cup) peas
130g (¾ cup) broad beans
handfuls of chopped mint, parsley and chives
85g (¾ cup) mature Cheddar cheese, grated
125ml (½ cup) milk
250ml (1 cup) double cream
2 whole eggs, plus 2 egg yolks

1 To make the pastry, chop then rub the butter into the flour and salt until the mixture resembles fine breadcrumbs. Add the egg yolk and mix in, and then slowly add a little milk to bring the mixture together into a ball. Wrap in cling film and refrigerate for 30 minutes.

2 Preheat the oven to 160°C/325°F/Gas 3. Roll out the pastry and line a loose-bottomed, greased 25cm (10 inch) tart tin with it. Prick the base with a fork, line it with baking parchment and ceramic baking beans or dried beans, then bake for 15 minutes. Remove the beans and baking parchment then return to the oven for a further 10 minutes. Seal the base by painting the egg white onto it and baking for a further 5 minutes.

3 Turn the oven up to 180°C/350°F/Gas 4. Prepare the filling by melting 15g (1 tablespoon) butter in a frying pan and gently frying the spring onions, whole, until they start to soften and colour up. Meanwhile, wilt the spinach in the remaining butter in a separate pan. Boil the peas and broad beans for 2 minutes then throw into ice-cold water. Arrange first the spinach, then herbs, then peas and beans, then spring onions in the pastry case and sprinkle with the grated cheese. Whisk together the milk, cream, eggs and egg yolks and pour over. Bake in the oven for 40 minutes. Serve either hot or cold.

Mange tout and sugar snap peas

80 days

Peas can be a tricky crop; even if they avoid the slugs and mice early on, there are maggots that tuck into the emerging peas later. Mange tout and sugar snap peas, on the other hand, are simple to grow. Not only just as good and with all that pea flavour, they are also quicker, as you don't have to wait for the peas to form: just cook crunchy pod and embryonic peas whole. You must also try the pea shoots: the little side shoots and growth tips are crunchy, sweet and pea-flavoured and look beautiful strewn on the top of a salad. You will get fewer pods if you pick them, but it's worth it if you have plenty of plants going. You can also sow peas into guttering or trays in autumn for a springtime crop of pea shoots.

Cultivation

Start mange tout and sugar snap peas off in pots under cover in early spring. They are pretty hardy so can go out into the ground in mid-spring, but the protection of a cloche will help them along until the weather really warms. They are climbers and like to scramble, so support them with sticks – shrubby offcuts (traditionally of hazel, but any deciduous prunings will do) pushed into the ground. A pea net stretched between two sturdy posts also makes the perfect scrambling surface. Once mange tout and sugar snap peas have made it past mice (pre-germination) and slugs (small seedling stage) they are fairly self-sufficient, requiring little else but watering during dry spells.

Recommended varieties

'Carouby de Maussanne' (mange tout) A delicious old French variety with gorgeous purple flowers followed by flat stringless peas.

'Sugar Ann' (sugar snap pea) Sweet and crunchy, best picked young.

'Shiraz' (mange tout) Dark purple pods that fade on cooking. Pick small and eat raw in colourful salads.

Harvesting and eating

Mange tout and sugar snap peas reward frequent picking; the more you pick, the more pods the plant will produce. The moment you let a pod swell to form full-sized seeds, the plant kicks back and stops producing more, so keep on the case. Eat raw in salads or lightly cooked. They are the perfect stir-fry ingredient – quick and crunchy – and also good lightly steamed.

Mark: "Peas like a deep root run. I grow them in root trainers, but if you can't find those, sow several seeds around the edges of a large pot. The cardboard cores inside toilet rolls can make good pea pots too."

Turnips

40 days

These dependable, distinctly unglamorous brassicas are easy to dismiss. They are unfairly considered bland and stodgy winter fillers, but if you choose early types and eat them as a sweet quick vegetable rather than growing them for bulk they are a tender and nutty revelation.

Cultivation

Turnips are hardy and easy. Sow them thinly, direct into warmed ground from early to late spring, and you'll be eating them all summer. They like a shady spot. The earliest sowings may try to go to seed prematurely, but keep on sowing every few weeks and you will have replacements. They are happier once the weather has warmed in summer, but make the final sowing in autumn for crops to eat through winter and early spring.

Turnips are brassicas, so they're troubled by flea beetles, which nibble holes in the leaves. Occasionally this can be severe enough that the plant fails to develop, but it usually isn't the case and they'll crop well as long as they get through the early stages without too much of a check. If it's causing you a real problem, try growing them under horticultural fleece. Turnips are also susceptible to club root, the other bane of brassicas, and as it lingers in the soil it's important to grow them in a different spot each year.

Recommended varieties

'Golden Globe' Very quick, with a tapered shape, golden colour and fine texture and flavour.

'Milan Purple Top' Delicious old variety that is pale in colour, with a purple top. It's a particularly good variety for autumn sowing.

'Petrowski' A golden-skinned turnip with a particularly sweet and mild flavour.

Turnips don't have to be stodgy winter fillers. Grown small and fast, they are a delight lightly roasted.

Harvesting and eating

Pick turnips when they are tiny for the best flavour. Often the root sits so close to the surface that you can just pull it out of the ground, but if not, loosen below it with a hand fork or trowel and then pull. They are at their best lightly roasted and thrown warm into a salad. The leafy tops are good, and spicy too: be sure to eat any of the thinnings and when you finally harvest the roots, make the tops the greens in the salad. Also try the roots sautéed in butter and garlic, whole if they are small enough or sliced if not, and the tops stir-fried or steamed. They make a great pasta sauce.

Turnip 'Milan Purple Top' and turnip 'Golden Globe'.

Honey and garlic roasted baby vegetables

3 handfuls of baby turnips, scrubbed
3 handfuls of baby carrots, scrubbed
3 handfuls of baby beetroot, halved
6 garlic cloves
olive oil
3 tablespoons runny honey

Preheat the oven to 200°C/400°F/Gas 6. Put the vegetables and garlic into a roasting tray and pour on the honey and a little olive oil, turning until all are coated. Roast for around 40 minutes until coloured and softened.

Beetroot/Beet

55 days

Beetroots are such stalwarts of the vegetable patch that it's easy to take them for granted. They are well loved for their natural sweetness and their earthy taste, and when picked early and tiny the sweetness comes to the fore, turning them from everyday to luxury. Don't confine yourself to the usual (though delicious) deep purple varieties; several of the differently coloured types are just as good, and far less messy to prepare.

Cultivation

Sow direct into well-prepared ground from early spring to mid- to late summer for harvesting from early summer into autumn. Make small sowings 2m (6½ feet) in length every few weeks to keep you in small beetroots throughout. They are fabulously untroubled by pests, so you can just sow and forget, thinning out as they germinate and grow to allow room for the roots to develop.

Recommended varieties

'Barbabietola di Chioggia' (also sold as 'Chioggia') Clear, concentric rings of candy pink and white are revealed when you cut it open. Worth growing for looks alone, but as impressive-tasting as any of the other varieties too.

'Bull's Blood' Tasty roots and beautiful deep red edible leaves. Pick and use both when tiny.

'Burpee's Golden' Fabulous egg-yolk yellow in colour, tasty, sweet and much less mess involved than with the purple kind.

'Sanguinea' The classic deep purple, sweet and juicy beetroot.

Mark: "Chard and beetroot come from the same plant, with one bred to be leafier, the other to produce a bulb. The leaves of beetroot have much of the flavour of chard – I pick them small and use them in a pasta sauce."

Harvesting and eating

Start harvesting when the roots are around golf-ball size or even smaller. The tiniest can just be washed and thrown raw into salads, leaves and all. Slightly larger ones will need cooking, but not too much. Wash off all the mud carefully, steam for 10 minutes or until tender, then slip off the skin. For an upmarket version of pickled beetroot, try scraping off the skin while raw (pretty easy when they're this small) and roast in olive oil and balsamic vinegar.

Little, golf-ball-sized beetroots 'Burpee's Golden' and 'Sanguinea'.

Chocolate and beetroot brownie

4 or 5 small beetroot
270g (2⅓ sticks) unsalted butter
255g (9 ounces) dark chocolate
3 eggs
200g (1 cup) caster (superfine) sugar
1 teaspoon vanilla extract
125g (1 cup) self-raising (self-rising) flour
pinch of sea salt

1 Boil the beetroot (skins and ends on) until tender (about 20–30 minutes). Plunge them into cold water and, wearing rubber gloves to stop your hands from staining, top, tail and peel them. Grate the beetroot into a bowl.

2 Preheat the oven to 180°C/350°F/Gas 4. Lightly grease and line a shallow 20 × 25cm (8 × 10 inch) baking tin with baking parchment. Chop the butter and chocolate into small pieces then place them in a bowl and sit it over a pan of simmering water, making sure the water doesn't touch the bottom of the bowl, to melt them gently.

3 Whisk the eggs, sugar and vanilla together, then beat in the melted chocolate and butter until smooth. Stir in the grated beetroot. Sift the flour and salt over the mixture and gently fold in with a large metal spoon.

4 Spoon the cake mix into the tin, smoothing the top with a spatula, and place in the oven for 20 minutes. Check if it's ready by inserting a knife – a few sticky crumbs should come out with the blade. If it's undercooked allow another 5–7 minutes cooking, but no more – a dry brownie is a disappointing thing. Once cooked, cool in the tin on a wire rack before cutting up.

Sources for seeds, plants and supplies

UK

Allotment Vegetable Growing
allotment.org.uk A source of useful equipment, vegetable seeds and advice.

Greenhouse Sensation
greenhousesensation.co.uk Suppliers of heated propagators.

Living Food of St Ives
livingfood.co.uk Sprouting seeds and equipment.

Otter Farm Shop
otterfarm.co.uk Specialists in the best-tasting varieties, with a good range of potatoes.

Real Seeds
realseeds.co.uk Breeders and collectors of the earliest producing varieties of vegetables.

Sarah Raven
sarahraven.com Particularly good range of seeds for cut-and-come-again salad leaves.

Sea Spring Seeds
seaspringseeds.co.uk Chilli pepper breeders with a great range of other vegetables as well, supplying seeds and young plants.

Simpsons Seeds
simpsonsseeds.co.uk Seeds and young plants of tomatoes, peppers and other greenhouse crops.

Slug Rings
slugrings.co.uk Solid copper rings to place around individual plants and protect them from slugs. Beautiful and effective.

The Organic Gardening Catalogue
organiccatalogue.com Seeds, organic fertilizers and pest and disease controls, containers, books and much more.

USA

Burpee
burpee.com Large selection of vegetable seeds including many heirlooms.

Johnny's Selected Seeds
johnnyseeds.com Great range of seeds for baby salad leaves.

Potato Garden
potatogarden.com Grow and sell a huge variety of seed potatoes.

Sprout People
sproutpeople.org Source of sprouting seeds and micro-green seeds, and sprouting equipment.

Canada

Seeds of Diversity Canada
www.seeds.ca

Annapolis Seeds
www.annapolisseeds.com

Hope Seed
www.hopeseed.com

Urban Harvest
www.uharvest.ca

Index

Mark Diacono runs the pioneering Otter Farm where he makes use of the changing climate to grow a wide range of food that is usually sourced from warmer climes. An award-winning journalist and photographer, Mark is also well known for his lectures, courses and work at River Cottage.

Lia Leendertz is a freelance journalist who writes for *The Guardian*, *Gardens Illustrated* and her award-winning blog, Midnight Brambling. She has written several other books, shares an allotment and studied horticulture at the Royal Botanic Gardens, Edinburgh.